Essentials of Geology

Essentials of Geology

Edited by
Collin Shephard

Larsen & Keller
www.larsen-keller.com

Essentials of Geology
Edited by Collin Shephard
ISBN: 978-1-63549-134-0 (Hardback)

© 2017 Larsen & Keller

▤ Larsen & Keller

Published by Larsen and Keller Education,
5 Penn Plaza,
19th Floor,
New York, NY 10001, USA

Cataloging-in-Publication Data

Essentials of geology / edited by Collin Shephard.
   p. cm.
Includes bibliographical references and index.
ISBN 978-1-63549-134-0
1. Geology. 2. Earth sciences. I. Shephard, Collin.
QE26.3 .E87 2017
550--dc23

The publisher's policy is to use permanent paper from mills that operate a sustainable forestry policy. Furthermore, the publisher ensures that the text paper and cover boards used have met acceptable environmental accreditation standards.

Printed and bound in the United States of America.

For more information regarding Larsen and Keller Education and its products, please visit the publisher's website www.larsen-keller.com

Table of Contents

Preface

Geology is an intricate field which deals with the study of solid surface of any celestial body, especially, Earth. It includes the examination of Earth's surface, its rocks and their morphology. It is important as it provides information about plate tectonics, past climate and also, the evolution of Earth. This important field is thoroughly explained in this extensive book. It provides deep insights into the subject and also explains its importance in the present day scenario. This text unfolds the complex aspects of geology which will be crucial for the holistic understanding of the subject matter. Some of the diverse topics covered in it address the varied branches that fall under this category. This textbook, with its detailed data, will prove immensely beneficial to professionals and students involved in this area at various levels.

To facilitate a deeper understanding of the contents of this book a short introduction of every chapter is written below:

Chapter 1- The study of solid Earth is known as geology. It studies the solid features of any astronomical object, such as the geology of the moon or Jupiter. Geological materials fall into two categories, rock and unconsolidated material. This chapter will provide an integrated understanding of geology.

Chapter 2- The majority of geological data comes from the study on solid Earth materials. Some of the branches of geology are petrology, mineralogy, structural geology and stratigraphy. Petrology studies the structure and composition of rocks whereas mineralogy is the subject of geology which deals with chemistry, crystal structure and minerals. This text is compilation of the various branches of geology that form an integral part of the broader subject matter.

Chapter 3- The principles of geology discussed within this section are crosscutting relationships, uniformitarianism, principle of original horizontality, law of superposition and principal of lateral continuity. One of the basic principles of geology is uniformitarianism; it is the supposition that same laws and processes that have in the past operated the universe, presently also operate it. This chapter elucidates the crucial theories and principles of geology.

Chapter 4- Sub-fields of geology include economic geology, engineering geology, mining, geotechnical engineering, plate tectonics, hydrogeology, planetary geology, biochemistry etc. Economic geology is the study of Earth's materials, mostly the ones that are used for economical or industrial purposes. This text helps the reader in developing an in-depth understanding of all the sub-fields of geology.

Chapter 5- Geology has a number of processes; some of these are shear, fold, saltation, metasomatism, denudation and spheroidal weathering. Geological fold occurs when the sedimentary strata becomes bent generally because of a permanent distortion. Saltation is another process related to geology that takes place with the help of wind or water, which help in the transportation of particles.

Chapter 6- Geological mapping is a vital part of geology; it is a special-purpose map that is created to illustrate geological features. The rocks or geological strata demonstrated in the map is shown by different colors and symbols. The major components of geology are discussed in the following text.

Chapter 7- Rocks are naturally formed aggregates of one or more minerals. The following text focuses on the formation of rocks and the rock cycle. The importance of rocks and their formations is a major element in the field of geology; the following text strategically encompasses and incorporates the major components and key concepts of rocks, providing a complete understanding.

Chapter 8- The history of geology concerns itself with the development that has taken place in the study of geology over a period of years. Geology studies the origin, history and structure of the earth. The section has been carefully written to provide an easy understanding of the history of geology.

Finally, I would like to thank the entire team involved since the inception of this book for their valuable time and contribution. This book would not have been possible without their efforts. I would also like to thank my friends and family for their constant support.

Editor

Introduction to Geology

The study of solid Earth is known as geology. It studies the solid features of any astronomical object, such as the geology of the moon or Jupiter. Geological materials fall into two categories, rock and unconsolidated material. This chapter will provide an integrated understanding of geology.

Geology is an earth science comprising the study of solid Earth, the rocks of which it is composed, and the processes by which they change. Geology can also refer generally to the study of the solid features of any celestial body (such as the geology of the Moon or Mars).

Geology gives insight into the history of the Earth by providing the primary evidence for plate tectonics, the evolutionary history of life, and past climates. Geology is important for mineral and hydrocarbon exploration and exploitation, evaluating water resources, understanding of natural hazards, the remediation of environmental problems, and for providing insights into past climate change. Geology also plays a role in geotechnical engineering and is a major academic discipline.

Geologic Materials

The majority of geological data comes from research on solid Earth materials. These typically fall into one of two categories: rock and unconsolidated material.

Rock

There are three major types of rock: igneous, sedimentary, and metamorphic. The rock cycle is an important concept in geology which illustrates the relationships between these three types of rock, and magma. When a rock crystallizes from melt (magma and/or lava), it is an igneous rock. This rock can be weathered and eroded, and then redeposited and lithified into a sedimentary rock, or be turned into a metamorphic rock due to heat and pressure that change the mineral content of the rock which gives it a characteristic fabric. The sedimentary rock can then be subsequently turned into a metamorphic rock due to heat and pressure and is then weathered, eroded, deposited, and lithified, ultimately becoming a sedimentary rock. Sedimentary rock may also be re-eroded and redeposited, and metamorphic rock may also undergo additional metamorphism. All three types of rocks may be re-melted; when this happens, a new magma is formed, from which an igneous rock may once again crystallize.

Rock Cycle

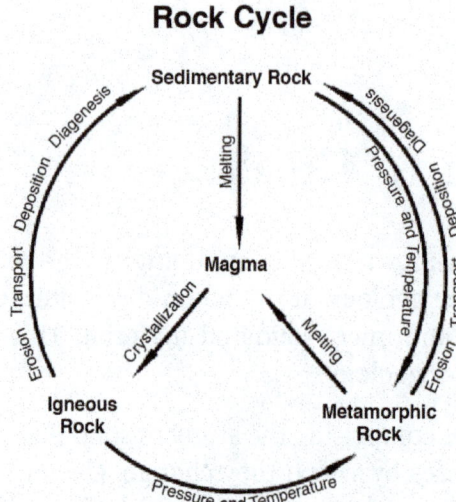

This schematic diagram of the rock cycle shows the relationship between magma and sedimentary, metamorphic, and igneous rock

The majority of research in geology is associated with the study of rock, as rock provides the primary record of the majority of the geologic history of the Earth.

Unconsolidated Material

Geologists also study unlithified material, which typically comes from more recent deposits. These materials are superficial deposits which lie above the bedrock. Because of this, the study of such material is often known as Quaternary geology, after the recent Quaternary Period. This includes the study of sediment and soils, including studies in geomorphology, sedimentology, and paleoclimatology.

Whole-Earth Structure

Plate Tectonics

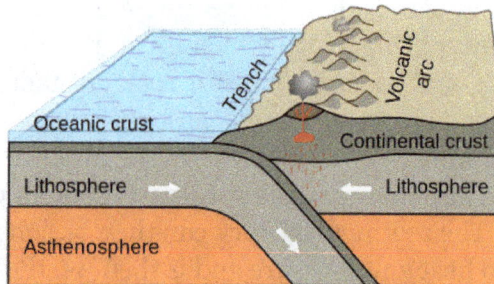

Oceanic-continental convergence resulting in subduction and volcanic arcs illustrates one effect of plate tectonics.

In the 1960s, a series of discoveries, the most important of which was seafloor spreading, showed that the Earth's lithosphere, which includes the crust and rigid uppermost portion of the upper mantle, is separated into a number of tectonic plates that move across the plastically deforming, solid, upper mantle, which is called the asthenosphere. There is an intimate coupling between the movement of the plates on the surface and the convection of the mantle: oceanic plate motions and mantle convection currents always move in the same direction, because the oceanic lithosphere is the rigid upper thermal boundary layer of the convecting mantle. This coupling between rigid plates moving on the surface of the Earth and the convecting mantle is called plate tectonics.

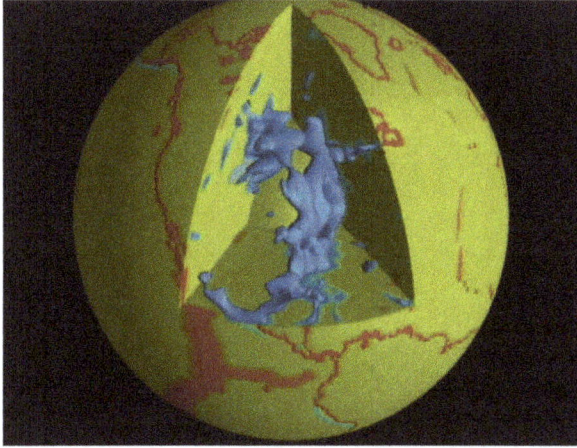

On this diagram, subducting slabs are in blue, and continental margins and a few plate boundaries are in red. The blue blob in the cutaway section is the seismically imaged Farallon Plate, which is subducting beneath North America. The remnants of this plate on the Surface of the Earth are the Juan de Fuca Plate and Explorer plate in the Northwestern USA / Southwestern Canada, and the Cocos Plate on the west coast of Mexico.

The development of plate tectonics provided a physical basis for many observations of the solid Earth. Long linear regions of geologic features could be explained as plate boundaries. Mid-ocean ridges, high regions on the seafloor where hydrothermal vents and volcanoes exist, were explained as divergent boundaries, where two plates move apart. Arcs of volcanoes and earthquakes were explained as convergent boundaries, where one plate subducts under another. Transform boundaries, such as the San Andreas Fault system, resulted in widespread powerful earthquakes. Plate tectonics also provided a mechanism for Alfred Wegener's theory of continental drift, in which the continents move across the surface of the Earth over geologic time. They also provided a driving force for crustal deformation, and a new setting for the observations of structural geology. The power of the theory of plate tectonics lies in its ability to combine all of these observations into a single theory of how the lithosphere moves over the convecting mantle.

Earth Structure

Advances in seismology, computer modeling, and mineralogy and crystallography at high temperatures and pressures give insights into the internal composition and structure of the Earth.

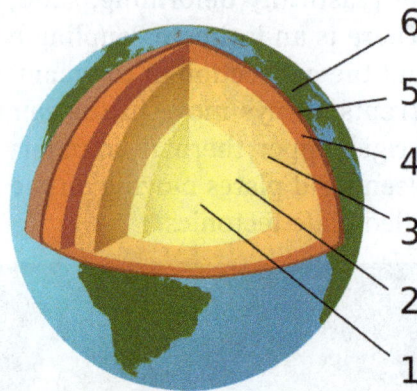

The Earth's layered structure. (1) inner core; (2) outer core; (3) lower mantle; (4) upper mantle; (5) lithosphere; (6) crust (part of the lithosphere)

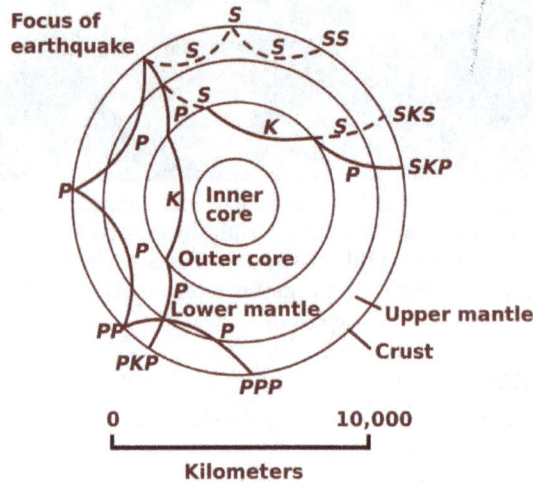

Earth layered structure. Typical wave paths from earthquakes like these gave early seismologists insights into the layered structure of the Earth

Seismologists can use the arrival times of seismic waves in reverse to image the interior of the Earth. Early advances in this field showed the existence of a liquid outer core (where shear waves were not able to propagate) and a dense solid inner core. These advances led to the development of a layered model of the Earth, with a crust and lithosphere on top, the mantle below (separated within itself by seismic discontinuities at 410 and 660 kilometers), and the outer core and inner core below that. More recently, seismologists have been able to create detailed images of wave speeds inside the earth in the same way a doctor images a body in a CT scan. These images have led to a much more detailed view of the interior of the Earth, and have replaced the simplified layered model with a much more dynamic model.

Mineralogists have been able to use the pressure and temperature data from the seismic and modelling studies alongside knowledge of the elemental composition of the

Earth to reproduce these conditions in experimental settings and measure changes in crystal structure. These studies explain the chemical changes associated with the major seismic discontinuities in the mantle and show the crystallographic structures expected in the inner core of the Earth.

Geologic Time

The geologic time scale encompasses the history of the Earth. It is bracketed at the old end by the dates of the earliest Solar System material at 4.567 Ga, (gigaannum: billion years ago) and the age of the Earth at 4.54 Ga at the beginning of the informally recognized Hadean eon. At the young end of the scale, it is bracketed by the present day in the Holocene epoch.

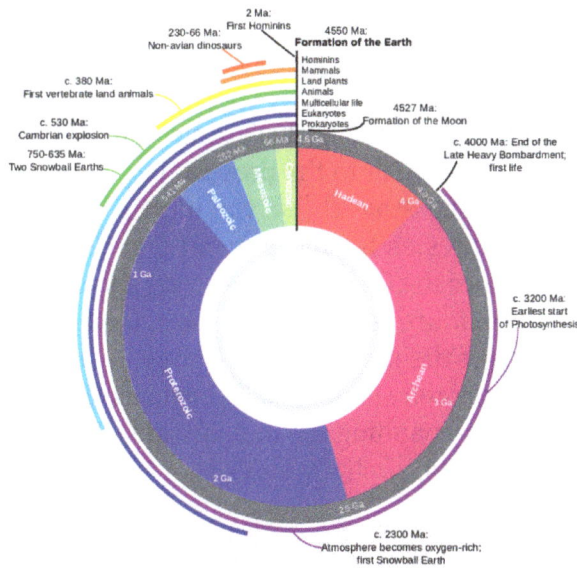

Geological time put in a diagram called a geological clock, showing the relative lengths of the eons of the Earth's history.

Important milestones

- 4.567 Ga: Solar system formation

- 4.54 Ga: Accretion of Earth

- c. 4 Ga: End of Late Heavy Bombardment, first life

- c. 3.5 Ga: Start of photosynthesis

- c. 2.3 Ga: Oxygenated atmosphere, first snowball Earth

- 730–635 Ma (megaannum: million years ago): second snowball Earth

- 542 ± 0.3 Ma: Cambrian explosion – vast multiplication of hard-bodied life; first abundant fossils; start of the Paleozoic

- c. 380 Ma: First vertebrate land animals

- 250 Ma: Permian-Triassic extinction – 90% of all land animals die; end of Paleozoic and beginning of Mesozoic

- 66 Ma: Cretaceous–Paleogene extinction – Dinosaurs die; end of Mesozoic and beginning of Cenozoic

- c. 7 Ma: First hominins appear

- 3.9 Ma: First Australopithecus, direct ancestor to modern Homo sapiens, appear

- 200 ka (kiloannum: thousand years ago): First modern Homo sapiens appear in East Africa

Brief Time Scale

The following four timelines show the geologic time scale. The first shows the entire time from the formation of the Earth to the present, but this compresses the most recent eon. Therefore, the second scale shows the most recent eon with an expanded scale. The second scale compresses the most recent era, so the most recent era is expanded in the third scale. Since the Quaternary is a very short period with short epochs, it is further expanded in the fourth scale. The second, third, and fourth timelines are therefore each subsections of their preceding timeline as indicated by asterisks. The Holocene (the latest epoch) is too small to be shown clearly on the third timeline on the right, another reason for expanding the fourth scale. The Pleistocene (P) epoch. Q stands for the Quaternary period.

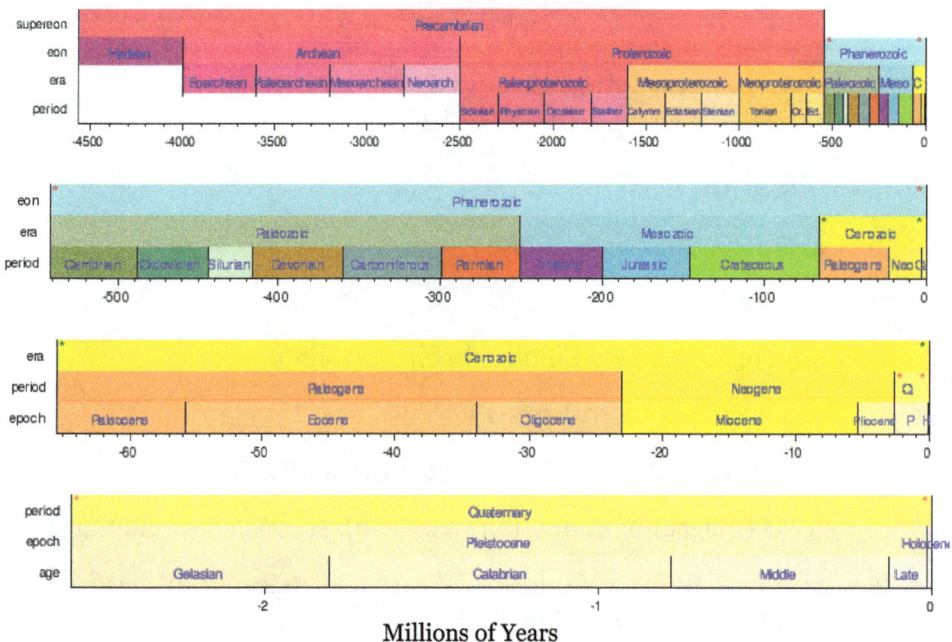

Millions of Years

Dating Methods

Geologists use a variety of methods to give both relative and absolute dates to geological events. They then use these dates to find the rates at which processes occur.

Relative Dating

Methods for relative dating were developed when geology first emerged as a formal science. Geologists still use the following principles today as a means to provide information about geologic history and the timing of geologic events.

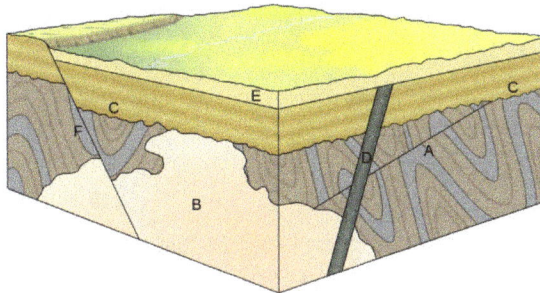

Cross-cutting relations can be used to determine the relative ages of rock strata and other geological structures. Explanations: A – folded rock strata cut by a thrust fault; B – large intrusion (cutting through A); C – erosional angular unconformity (cutting off A & B) on which rock strata were deposited; D – volcanic dyke (cutting through A, B & C); E – even younger rock strata (overlying C & D); F – normal fault (cutting through A, B, C & E).

The principle of Uniformitarianism states that the geologic processes observed in operation that modify the Earth's crust at present have worked in much the same way over geologic time. A fundamental principle of geology advanced by the 18th century Scottish physician and geologist James Hutton, is that "the present is the key to the past." In Hutton's words: "the past history of our globe must be explained by what can be seen to be happening now."

The principle of intrusive relationships concerns crosscutting intrusions. In geology, when an igneous intrusion cuts across a formation of sedimentary rock, it can be determined that the igneous intrusion is younger than the sedimentary rock. There are a number of different types of intrusions, including stocks, laccoliths, batholiths, sills and dikes.

The principle of cross-cutting relationships pertains to the formation of faults and the age of the sequences through which they cut. Faults are younger than the rocks they cut; accordingly, if a fault is found that penetrates some formations but not those on top of it, then the formations that were cut are older than the fault, and the ones that are not cut must be younger than the fault. Finding the key bed in these situations may help determine whether the fault is a normal fault or a thrust fault.

The principle of inclusions and components states that, with sedimentary rocks, if inclusions (or *clasts*) are found in a formation, then the inclusions must be older than the formation that contains them. For example, in sedimentary rocks, it is common for gravel from an older formation to be ripped up and included in a newer layer. A similar situation with igneous rocks occurs when xenoliths are found. These foreign bodies are picked up as magma or lava flows, and are incorporated, later to cool in the matrix. As a result, xenoliths are older than the rock which contains them.

The Permian through Jurassic stratigraphy of the Colorado Plateau area of southeastern Utah is a great example of both Original Horizontality and the Law of Superposition. These strata make up much of the famous prominent rock formations in widely spaced protected areas such as Capitol Reef National Park and Canyonlands National Park. From top to bottom: Rounded tan domes of the Navajo Sandstone, layered red Kayenta Formation, cliff-forming, vertically jointed, red Wingate Sandstone, slope-forming, purplish Chinle Formation, layered, lighter-red Moenkopi Formation, and white, layered Cutler Formation sandstone. Picture from Glen Canyon National Recreation Area, Utah.

The principle of original horizontality states that the deposition of sediments occurs as essentially horizontal beds. Observation of modern marine and non-marine sediments in a wide variety of environments supports this generalization (although cross-bedding is inclined, the overall orientation of cross-bedded units is horizontal).

The principle of superposition states that a sedimentary rock layer in a tectonically undisturbed sequence is younger than the one beneath it and older than the one above it. Logically a younger layer cannot slip beneath a layer previously deposited. This principle allows sedimentary layers to be viewed as a form of vertical time line, a partial or complete record of the time elapsed from deposition of the lowest layer to deposition of the highest bed.

The principle of faunal succession is based on the appearance of fossils in sedimentary rocks. As organisms exist at the same time period throughout the world, their presence or (sometimes) absence may be used to provide a relative age of the formations in which they are found. Based on principles laid out by William Smith almost a hundred years before the publication of Charles Darwin's theory of evolution, the principles of succession were developed independently of evolutionary thought. The principle becomes quite complex, however, given the uncertainties of fossilization, the localization of fossil types due to lateral changes in habitat (facies change in sedimentary strata), and that not all fossils may be found globally at the same time.

Absolute Dating

Geologists also use methods to determine the absolute age of rock samples and geological events. These dates are useful on their own and may also be used in conjunction with relative dating methods or to calibrate relative methods.

At the beginning of the 20th century, important advancement in geological science was facilitated by the ability to obtain accurate absolute dates to geologic events using radioactive isotopes and other methods. This changed the understanding of geologic time. Previously, geologists could only use fossils and stratigraphic correlation to date sections of rock relative to one another. With isotopic dates it became possible to assign absolute ages to rock units, and these absolute dates could be applied to fossil sequences in which there was datable material, converting the old relative ages into new absolute ages.

For many geologic applications, isotope ratios of radioactive elements are measured in minerals that give the amount of time that has passed since a rock passed through its particular closure temperature, the point at which different radiometric isotopes stop diffusing into and out of the crystal lattice. These are used in geochronologic and thermochronologic studies. Common methods include uranium-lead dating, potassium-argon dating, argon-argon dating and uranium-thorium dating. These methods are used for a variety of applications. Dating of lava and volcanic ash layers found within a stratigraphic sequence can provide absolute age data for sedimentary rock units which do not contain radioactive isotopes and calibrate relative dating techniques. These methods can also be used to determine ages of pluton emplacement. Thermochemical techniques can be used to determine temperature profiles within the crust, the uplift of mountain ranges, and paleotopography.

Fractionation of the lanthanide series elements is used to compute ages since rocks were removed from the mantle.

Other methods are used for more recent events. Optically stimulated luminescence and cosmogenic radionucleide dating arc uscd to datc surfaccs and/or erosion rates. Dendrochronology can also be used for the dating of landscapes. Radiocarbon dating is used for geologically young materials containing organic carbon.

Geological Development of an Area

The geology of an area changes through time as rock units are deposited and inserted and deformational processes change their shapes and locations.

Rock units are first emplaced either by deposition onto the surface or intrusion into the overlying rock. Deposition can occur when sediments settle onto the surface of the Earth and later lithify into sedimentary rock, or when as volcanic material such as volcanic ash or lava flows blanket the surface. Igneous intrusions such as batholiths, laccoliths, dikes, and sills, push upwards into the overlying rock, and crystallize as they intrude.

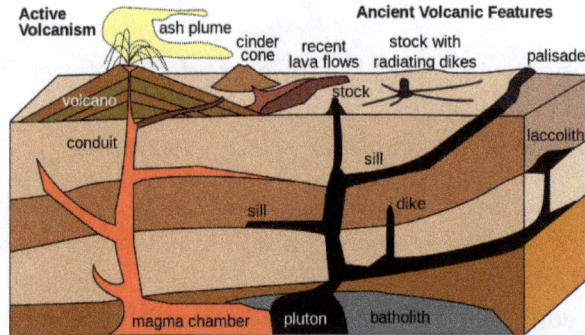

An originally horizontal sequence of sedimentary rocks (in shades of tan) are affected by igneous activity. Deep below the surface are a magma chamber and large associated igneous bodies. The magma chamber feeds the volcano, and sends offshoots of magma that will later crystallize into dikes and sills. Magma also advances upwards to form intrusive igneous bodies. The diagram illustrates both a cinder cone volcano, which releases ash, and a composite volcano, which releases both lava and ash.

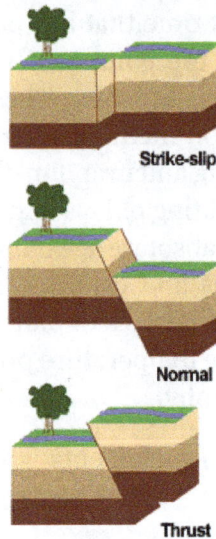

An illustration of the three types of faults. Strike-slip faults occur when rock units slide past one another, normal faults occur when rocks are undergoing horizontal extension, and thrust faults occur when rocks are undergoing horizontal shortening.

After the initial sequence of rocks has been deposited, the rock units can be deformed and/or metamorphosed. Deformation typically occurs as a result of horizontal shortening, horizontal extension, or side-to-side (strike-slip) motion. These structural regimes broadly relate to convergent boundaries, divergent boundaries, and transform boundaries, respectively, between tectonic plates.

When rock units are placed under horizontal compression, they shorten and become thicker. Because rock units, other than muds, do not significantly change in volume, this is accomplished in two primary ways: through faulting and folding. In the shallow crust, where brittle deformation can occur, thrust faults form, which cause deeper rock to move on top of shallower rock. Because deeper rock is often older, as noted by the principle of superposition, this can result in older rocks moving on top of younger ones. Movement along faults can result in folding, either because the faults are not planar or because rock layers are dragged along, forming drag folds as slip occurs along the fault. Deeper in the Earth, rocks behave plastically, and fold instead of faulting. These folds can either be those where the material in the center of the fold buckles upwards, creating "antiforms", or where it buckles downwards, creating "synforms". If the tops of the rock units within the folds remain pointing upwards, they are called anticlines and synclines, respectively. If some of the units in the fold are facing downward, the structure is called an overturned anticline or syncline, and if all of the rock units are overturned or the correct up-direction is unknown, they are simply called by the most general terms, antiforms and synforms.

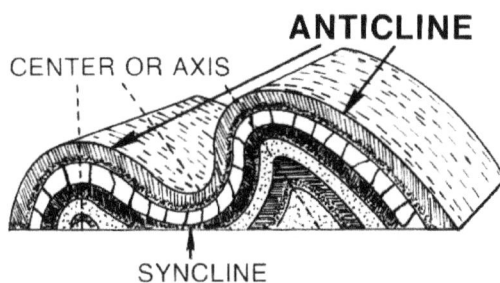

A diagram of folds, indicating an anticline and a syncline.

Even higher pressures and temperatures during horizontal shortening can cause both folding and metamorphism of the rocks. This metamorphism causes changes in the mineral composition of the rocks; creates a foliation, or planar surface, that is related to mineral growth under stress. This can remove signs of the original textures of the rocks, such as bedding in sedimentary rocks, flow features of lavas, and crystal patterns in crystalline rocks.

Extension causes the rock units as a whole to become longer and thinner. This is primarily accomplished through normal faulting and through the ductile stretching and thinning. Normal faults drop rock units that are higher below those that are lower. This typically results in younger units being placed below older units. Stretching of units can result in their thinning; in fact, there is a location within the Maria Fold and Thrust Belt in which the entire sedimentary sequence of the Grand Canyon can be seen over a length of less than a meter. Rocks at the depth to be ductilely stretched are often also metamorphosed. These stretched rocks can also pinch into lenses, known as boudins, after the French word for "sausage", because of their visual similarity.

Where rock units slide past one another, strike-slip faults develop in shallow regions, and become shear zones at deeper depths where the rocks deform ductilely.

The addition of new rock units, both depositionally and intrusively, often occurs during deformation. Faulting and other deformational processes result in the creation of topographic gradients, causing material on the rock unit that is increasing in elevation to be eroded by hillslopes and channels. These sediments are deposited on the rock unit that is going down. Continual motion along the fault maintains the topographic gradient in spite of the movement of sediment, and continues to create accommodation space for the material to deposit. Deformational events are often also associated with volcanism and igneous activity. Volcanic ashes and lavas accumulate on the surface, and igneous intrusions enter from below. Dikes, long, planar igneous intrusions, enter along cracks, and therefore often form in large numbers in areas that are being actively deformed. This can result in the emplacement of dike swarms, such as those that are observable across the Canadian shield, or rings of dikes around the lava tube of a volcano.

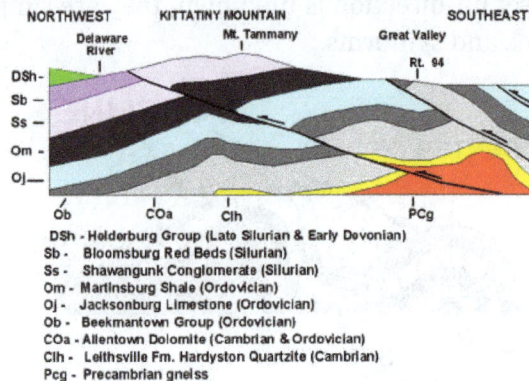

DSh - Helderburg Group (Late Silurian & Early Devonian)
Sb - Bloomsburg Red Beds (Silurian)
Ss - Shawangunk Conglomerate (Silurian)
Om - Martinsburg Shale (Ordovician)
Oj - Jacksonburg Limestone (Ordovician)
Ob - Beekmantown Group (Ordovician)
COa - Allentown Dolomite (Cambrian & Ordovician)
Clh - Leithsville Fm. Hardyston Quartzite (Cambrian)
Pcg - Precambrian gneiss

Geologic cross section of Kittatinny Mountain. This cross section shows metamorphic rocks, overlain by younger sediments deposited after the metamorphic event. These rock units were later folded and faulted during the uplift of the mountain.

All of these processes do not necessarily occur in a single environment, and do not necessarily occur in a single order. The Hawaiian Islands, for example, consist almost entirely of layered basaltic lava flows. The sedimentary sequences of the mid-continental United States and the Grand Canyon in the southwestern United States contain almost-undeformed stacks of sedimentary rocks that have remained in place since Cambrian time. Other areas are much more geologically complex. In the southwestern United States, sedimentary, volcanic, and intrusive rocks have been metamorphosed, faulted, foliated, and folded. Even older rocks, such as the Acasta gneiss of the Slave craton in northwestern Canada, the oldest known rock in the world have been metamorphosed to the point where their origin is undiscernable without laboratory analysis. In addition, these processes can occur in stages. In many places, the Grand Canyon in the southwestern United States being a very visible example, the lower rock units were metamorphosed and deformed, and then deformation ended and the upper, undeformed units were deposited. Although any amount of rock emplacement and rock deformation can occur, and they can occur any number of times, these concepts provide a guide to understanding the geological history of an area.

Methods of Geology

Geologists use a number of field, laboratory, and numerical modeling methods to decipher Earth history and understand the processes that occur on and inside the Earth. In typical geological investigations, geologists use primary information related to petrology (the study of rocks), stratigraphy (the study of sedimentary layers), and structural geology (the study of positions of rock units and their deformation). In many cases, geologists also study modern soils, rivers, landscapes, and glaciers; investigate past and current life and biogeochemical pathways, and use geophysical methods to investigate the subsurface.

Field Methods

A standard Brunton Pocket Transit, commonly used by geologists for mapping and surveying.

A typical USGS field mapping camp in the 1950s

Today, handheld computers with GPS and geographic information systems software are often used in geological field work (digital geologic mapping).

Geological field work varies depending on the task at hand. Typical fieldwork could consist of:

- Geological mapping

- o   Structural mapping: the locations of major rock units and the faults and folds that led to their placement there.

- o   Stratigraphic mapping: the locations of sedimentary facies (lithofacies and biofacies) or the mapping of isopachs of equal thickness of sedimentary rock

- o   Surficial mapping: the locations of soils and surficial deposits

- Surveying of topographic features

    - o   Creation of topographic maps

    - o   Work to understand change across landscapes, including:

        - ▪   Patterns of erosion and deposition

        - ▪   River channel change through migration and avulsion

        - ▪   Hillslope processes

- Subsurface mapping through geophysical methods

    - o   These methods include:

        - ▪   Shallow seismic surveys

        - ▪   Ground-penetrating radar

        - ▪   Aeromagnetic surveys

        - ▪   Electrical resistivity tomography

    - o   They are used for:

        - ▪   Hydrocarbon exploration

        - ▪   Finding groundwater

        - ▪   Locating buried archaeological artifacts

- High-resolution stratigraphy

    - o   Measuring and describing stratigraphic sections on the surface

    - o   Well drilling and logging

- Biogeochemistry and geomicrobiology

    - o   Collecting samples to:

        - ▪   Determine biochemical pathways

        - ▪   Identify new species of organisms

- ▪ Identify new chemical compounds
  - o And to use these discoveries to:
    - ▪ Understand early life on Earth and how it functioned and metabolized
    - ▪ Find important compounds for use in pharmaceuticals.
- Paleontology: excavation of fossil material
  - o For research into past life and evolution
  - o For museums and education
- Collection of samples for geochronology and thermochronology
- Glaciology: measurement of characteristics of glaciers and their motion

Petrology

In addition to identifying rocks in the field, petrologists identify rock samples in the laboratory. Two of the primary methods for identifying rocks in the laboratory are through optical microscopy and by using an electron microprobe. In an optical mineralogy analysis, thin sections of rock samples are analyzed through a petrographic microscope, where the minerals can be identified through their different properties in plane-polarized and cross-polarized light, including their birefringence, pleochroism, twinning, and interference properties with a conoscopic lens. In the electron microprobe, individual locations are analyzed for their exact chemical compositions and variation in composition within individual crystals. Stable and radioactive isotope studies provide insight into the geochemical evolution of rock units.

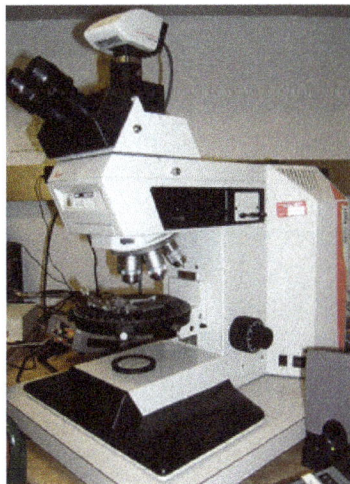

A petrographic microscope, which is an optical microscope fitted with cross-polarizing lenses, a conoscopic lens, and compensators (plates of anisotropic materials; gypsum plates and quartz wedges are common), for crystallographic analysis.

Petrologists can also use fluid inclusion data and perform high temperature and pressure physical experiments to understand the temperatures and pressures at which different mineral phases appear, and how they change through igneous and metamorphic processes. This research can be extrapolated to the field to understand metamorphic processes and the conditions of crystallization of igneous rocks. This work can also help to explain processes that occur within the Earth, such as subduction and magma chamber evolution.

Structural Geology

Structural geologists use microscopic analysis of oriented thin sections of geologic samples to observe the fabric within the rocks which gives information about strain within the crystalline structure of the rocks. They also plot and combine measurements of geological structures in order to better understand the orientations of faults and folds in order to reconstruct the history of rock deformation in the area. In addition, they perform analog and numerical experiments of rock deformation in large and small settings.

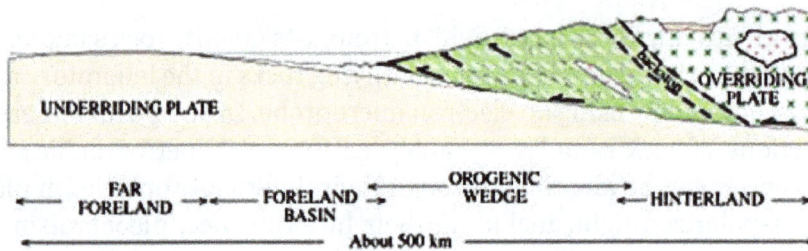

A diagram of an orogenic wedge. The wedge grows through faulting in the interior and along the main basal fault, called the décollement. It builds its shape into a critical taper, in which the angles within the wedge remain the same as failures inside the material balance failures along the décollement. It is analogous to a bulldozer pushing a pile of dirt, where the bulldozer is the overriding plate.

The analysis of structures is often accomplished by plotting the orientations of various features onto stereonets. A stereonet is a stereographic projection of a sphere onto a plane, in which planes are projected as lines and lines are projected as points. These can be used to find the locations of fold axes, relationships between faults, and relationships between other geologic structures.

Among the most well-known experiments in structural geology are those involving orogenic wedges, which are zones in which mountains are built along convergent tectonic plate boundaries. In the analog versions of these experiments, horizontal layers of sand are pulled along a lower surface into a back stop, which results in realistic-looking patterns of faulting and the growth of a critically tapered (all angles remain the same) orogenic wedge. Numerical models work in the same way as these analog models, though they are often more sophisticated and can include patterns of erosion and uplift in the mountain belt. This helps to show the relationship between erosion and the shape of the mountain range. These studies can also give useful information about pathways for metamorphism through pressure, temperature, space, and time.

Stratigraphy

In the laboratory, stratigraphers analyze samples of stratigraphic sections that can be returned from the field, such as those from drill cores. Stratigraphers also analyze data from geophysical surveys that show the locations of stratigraphic units in the subsurface. Geophysical data and well logs can be combined to produce a better view of the subsurface, and stratigraphers often use computer programs to do this in three dimensions. Stratigraphers can then use these data to reconstruct ancient processes occurring on the surface of the Earth, interpret past environments, and locate areas for water, coal, and hydrocarbon extraction.

In the laboratory, biostratigraphers analyze rock samples from outcrop and drill cores for the fossils found in them. These fossils help scientists to date the core and to understand the depositional environment in which the rock units formed. Geochronologists precisely date rocks within the stratigraphic section in order to provide better absolute bounds on the timing and rates of deposition. Magnetic stratigraphers look for signs of magnetic reversals in igneous rock units within the drill cores. Other scientists perform stable isotope studies on the rocks to gain information about past climate.

Planetary Geology

With the advent of space exploration in the twentieth century, geologists have begun to look at other planetary bodies in the same ways that have been developed to study the Earth. This new field of study is called planetary geology (sometimes known as astrogeology) and relies on known geologic principles to study other bodies of the solar system.

Surface of Mars as photographed by the Viking 2 lander December 9, 1977.

Although the Greek-language-origin prefix *geo* refers to Earth, "geology" is often used in conjunction with the names of other planetary bodies when describing their composition and internal processes: examples are "the geology of Mars" and "Lunar geology". Specialised terms such as *selenology* (studies of the Moon), *areology* (of Mars), etc., are also in use.

Although planetary geologists are interested in studying all aspects of other planets, a significant focus is to search for evidence of past or present life on other worlds. This has led to many missions whose primary or ancillary purpose is to examine planetary bodies for evidence of life. One of these is the Phoenix lander, which analyzed Martian polar soil for water, chemical, and mineralogical constituents related to biological processes.

Applied Geology

Economic Geology

Economic geology is an important branch of geology which deals with different aspects of economic minerals being used by humankind to fulfill its various needs. The economic minerals are those which can be extracted profitably. Economic geologists help locate and manage the Earth's natural resources, such as petroleum and coal, as well as mineral resources, which include metals such as iron, copper, and uranium.

Mining Geology

Mining geology consists of the extractions of mineral resources from the Earth. Some resources of economic interests include gemstones, metals, and many minerals such as asbestos, perlite, mica, phosphates, zeolites, clay, pumice, quartz, and silica, as well as elements such as sulfur, chlorine, and helium.

Petroleum Geology

Petroleum geologists study the locations of the subsurface of the Earth which can contain extractable hydrocarbons, especially petroleum and natural gas. Because many of these reservoirs are found in sedimentary basins, they study the formation of these basins, as well as their sedimentary and tectonic evolution and the present-day positions of the rock units.

Mud log in process, a common way to study the lithology when drilling oil wells.

Engineering Geology

Engineering geology is the application of the geologic principles to engineering practice for the purpose of assuring that the geologic factors affecting the location, design, construction, operation, and maintenance of engineering works are properly addressed.

In the field of civil engineering, geological principles and analyses are used in order to ascertain the mechanical principles of the material on which structures are built. This allows tunnels to be built without collapsing, bridges and skyscrapers to be built with sturdy foundations, and buildings to be built that will not settle in clay and mud.

Hydrology and Environmental Issues

Geology and geologic principles can be applied to various environmental problems such as stream restoration, the restoration of brownfields, and the understanding of the interaction between natural habitat and the geologic environment. Groundwater hydrology, or hydrogeology, is used to locate groundwater, which can often provide a ready supply of uncontaminated water and is especially important in arid regions, and to monitor the spread of contaminants in groundwater wells.

Geologists also obtain data through stratigraphy, boreholes, core samples, and ice cores. Ice cores and sediment cores are used to for paleoclimate reconstructions, which tell geologists about past and present temperature, precipitation, and sea level across the globe. These datasets are our primary source of information on global climate change outside of instrumental data.

Natural Hazards

Geologists and geophysicists study natural hazards in order to enact safe building codes and warning systems that are used to prevent loss of property and life. Examples of important natural hazards that are pertinent to geology (as opposed those that are mainly or only pertinent to meteorology) are:

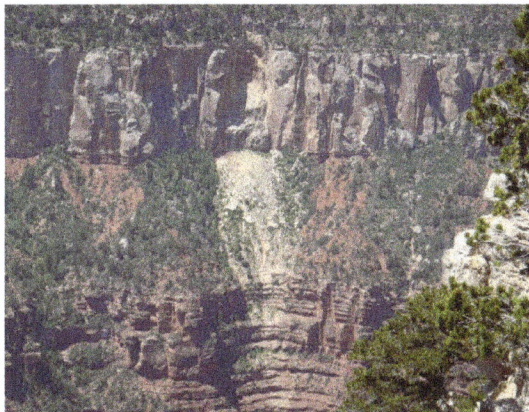

Rockfall in the Grand Canyon

- Avalanches

- Earthquakes

- Floods

- Landslides and debris flows

- River channel migration and avulsion

- Liquefaction

- Sinkholes

- Subsidence

- Tsunamis

- Volcanoes

History of Geology

The study of the physical material of the Earth dates back at least to ancient Greece when Theophrastus (372–287 BCE) wrote the work *Peri Lithon* (*On Stones*). During the Roman period, Pliny the Elder wrote in detail of the many minerals and metals then in practical use – even correctly noting the origin of amber.

William Smith's geologic map of England, Wales, and southern Scotland. Completed in 1815, it was the second national-scale geologic map, and by far the most accurate of its time.

Some modern scholars, such as Fielding H. Garrison, are of the opinion that the origin of the science of geology can be traced to Persia after the Muslim conquests had come to an end. Abu al-Rayhan al-Biruni (973–1048 CE) was one of the earliest Persian geologists, whose works included the earliest writings on the geology of India, hypothesizing that the Indian subcontinent was once a sea. Drawing from Greek and Indian scientific literature that were not destroyed by the Muslim conquests, the Persian scholar Ibn Sina (Avicenna, 981–1037) proposed detailed explanations for the formation of mountains, the origin of earthquakes, and other topics central to modern geology, which provided an essential foundation for the later development of the science. In China, the polymath Shen Kuo (1031–1095) formulated a hypothesis for the process of land formation: based on his observation of fossil animal shells in a geological stratum in a mountain hundreds of miles from the ocean, he inferred that the land was formed by erosion of the mountains and by deposition of silt.

Nicolas Steno (1638–1686) is credited with the law of superposition, the principle of original horizontality, and the principle of lateral continuity: three defining principles of stratigraphy.

The word *geology* was first used by Ulisse Aldrovandi in 1603, then by Jean-André Deluc in 1778 and introduced as a fixed term by Horace-Bénédict de Saussure in 1779. The word is derived from the Greek γ□, *gê*, meaning "earth" and λόγος, *logos*, meaning "speech". But according to another source, the word "geology" comes from a Norwegian, Mikkel Pedersøn Escholt (1600–1699), who was a priest and scholar. Escholt first used the definition in his book titled, *Geologica Norvegica* (1657).

William Smith (1769–1839) drew some of the first geological maps and began the process of ordering rock strata (layers) by examining the fossils contained in them.

James Hutton is often viewed as the first modern geologist. In 1785 he presented a paper entitled *Theory of the Earth* to the Royal Society of Edinburgh. In his paper, he explained his theory that the Earth must be much older than had previously been supposed in order to allow enough time for mountains to be eroded and for sediments to form new rocks at the bottom of the sea, which in turn were raised up to become dry land. Hutton published a two-volume version of his ideas in 1795 (Vol. 1, Vol. 2).

Scotsman James Hutton, father of modern geology

Followers of Hutton were known as *Plutonists* because they believed that some rocks were formed by *vulcanism*, which is the deposition of lava from volcanoes, as opposed to the *Neptunists*, led by Abraham Werner, who believed that all rocks had settled out of a large ocean whose level gradually dropped over time.

The first geological map of the U.S. was produced in 1809 by William Maclure. In 1807, Maclure commenced the self-imposed task of making a geological survey of the United States. Almost every state in the Union was traversed and mapped by him, the Allegheny Mountains being crossed and recrossed some 50 times. The results of his unaided labours were submitted to the American Philosophical Society in a memoir entitled *Observations on the Geology of the United States explanatory of a Geological Map*, and published in the *Society's Transactions*, together with the nation's first geological map. This antedates William Smith's geological map of England by six years, although it was constructed using a different classification of rocks.

Sir Charles Lyell first published his famous book, *Principles of Geology*, in 1830. This book, which influenced the thought of Charles Darwin, successfully promoted the doctrine of uniformitarianism. This theory states that slow geological processes have occurred throughout the Earth's history and are still occurring today. In contrast, catastrophism is the theory that Earth's features formed in single, catastrophic events and remained unchanged thereafter. Though Hutton believed in uniformitarianism, the idea was not widely accepted at the time.

Much of 19th-century geology revolved around the question of the Earth's exact age. Estimates varied from a few hundred thousand to billions of years. By the early 20th century, radiometric dating allowed the Earth's age to be estimated at two billion years. The awareness of this vast amount of time opened the door to new theories about the processes that shaped the planet.

Some of the most significant advances in 20th-century geology have been the development of the theory of plate tectonics in the 1960s and the refinement of estimates of the planet's age. Plate tectonics theory arose from two separate geological observations: seafloor spreading and continental drift. The theory revolutionized the Earth sciences. Today the Earth is known to be approximately 4.5 billion years old.

References

- Rollinson, Hugh R. (1996). Using geochemical data evaluation, presentation, interpretation. Harlow: Longman. ISBN 978-0-582-06701-1.

- Faure, Gunter (1998). Principles and applications of geochemistry: a comprehensive textbook for geology students. Upper Saddle River, NJ: Prentice-Hall. ISBN 978-0-02-336450-1.

- Burger, H. Robert; Sheehan, Anne F.; Jones, Craig H. (2006). Introduction to applied geophysics : exploring the shallow subsurface. New York: W.W. Norton. ISBN 0-393-92637-0.

- Krumbein, Wolfgang E., ed. (1978). Environmental biogeochemistry and geomicrobiology. Ann Arbor, Mich.: Ann Arbor Science Publ. ISBN 0-250-40218-1.

- Hubbard, Bryn; Glasser, Neil (2005). Field techniques in glaciology and glacial geomorphology. Chichester, England: J. Wiley. ISBN 0-470-84426-4.

- Nesse, William D. (1991). Introduction to optical mineralogy. New York: Oxford University Press. ISBN 0-19-506024-5.

- Shepherd, T.J.; Rankin, A.H.; Alderton, D.H.M. (1985). A practical guide to fluid inclusion studies. Glasgow: Blackie. ISBN 0-412-00601-4.

- McBirney, Alexander R. (2007). Igneous petrology. Boston: Jones and Bartlett Publishers. ISBN 978-0-7637-3448-0.

- Spear, Frank S. (1995). Metamorphic phase equilibria and pressure-temperature-time paths. Washington, DC: Mineralogical Soc. of America. ISBN 978-0-939950-34-8.

- Bally, A.W., ed. (1987). Atlas of seismic stratigraphy. Tulsa, Okla., U.S.A.: American Association of Petroleum Geologists. ISBN 0-89181-033-1.

- Winchester, Simon (2002). The map that changed the world: William Smith and the birth of modern geology. New York, NY: Perennial. ISBN 0-06-093180-9.

2

Branches of Geology

The majority of geological data comes from the study on solid Earth materials. Some of the branches of geology are petrology, mineralogy, structural geology and stratigraphy. Petrology studies the structure and composition of rocks whereas mineralogy is the subject of geology which deals with chemistry, crystal structure and minerals. This text is compilation of the various branches of geology that form an integral part of the broader subject matter.

Petrology

Petrology is the branch of geology that studies the origin, composition, distribution and structure of rocks.

A volcanic sand grain seen under the microscope, with plane-polarized light in the upper picture, and cross polarized light in the lower picture. Scale box is 0.25 mm.

Lithology was once approximately synonymous with petrography, but in current usage, lithology focuses on macroscopic hand-sample or outcrop-scale description of rocks while petrography is the speciality that deals with microscopic details.

In the petroleum industry, lithology, or more specifically mud logging, is the graphic representation of geological formations being drilled through, and drawn on a log called a mud log. As the cuttings are circulated out of the borehole they are sampled, examined (typically under a 10× microscope) and tested chemically when needed.

Methodology

Petrology utilizes the fields of mineralogy, petrography, optical mineralogy, and chemical analysis to describe the composition and texture of rocks. Petrologists also include the principles of geochemistry and geophysics through the study of geochemical trends and cycles and the use of thermodynamic data and experiments in order to better understand the origins of rocks.

Branches

There are three branches of petrology, corresponding to the three types of rocks: igneous, metamorphic, and sedimentary, and another dealing with experimental techniques:

- Igneous petrology focuses on the composition and texture of igneous rocks (rocks such as granite or basalt which have crystallized from molten rock or magma). Igneous rocks include volcanic and plutonic rocks.

- Sedimentary petrology focuses on the composition and texture of sedimentary rocks (rocks such as sandstone, shale, or limestone which consist of pieces or particles derived from other rocks or biological or chemical deposits, and are usually bound together in a matrix of finer material).

- Metamorphic petrology focuses on the composition and texture of metamorphic rocks (rocks such as slate, marble, gneiss, or schist which started out as sedimentary or igneous rocks but which have undergone chemical, mineralogical or textural changes due to extremes of pressure, temperature or both)

- Experimental petrology employs high-pressure, high-temperature apparatus to investigate the geochemistry and phase relations of natural or synthetic materials at elevated pressures and temperatures. Experiments are particularly useful for investigating rocks of the lower crust and upper mantle that rarely survive the journey to the surface in pristine condition. They are also one of the prime sources of information about completely inaccessible rocks such as those in the Earth's lower mantle and in the mantles of the other terrestrial planets and the Moon. The work of experimental petrologists has laid a foundation on which modern understanding of igneous and metamorphic processes has been built.

Petrography

Petrography is a branch of petrology that focuses on detailed descriptions of rocks.

Someone who studies petrography is called a petrographer. The mineral content and the textural relationships within the rock are described in detail. The classification of rocks is based on the information acquired during the petrographic analysis. Petrographic descriptions start with the field notes at the outcrop and include macroscopic description of hand specimens. However, the most important tool for the petrographer is the petrographic microscope. The detailed analysis of minerals by optical mineralogy in thin section and the micro-texture and structure are critical to understanding the origin of the rock. Electron microprobe analysis of individual grains as well as whole rock chemical analysis by atomic absorption, X-ray fluorescence, and laser-induced breakdown spectroscopy are used in a modern petrographic lab. Individual mineral grains from a rock sample may also be analyzed by X-ray diffraction when optical means are insufficient. Analysis of microscopic fluid inclusions within mineral grains with a heating stage on a petrographic microscope provides clues to the temperature and pressure conditions existent during the mineral formation.

History

Petrography as a science began in 1828 when Scottish physicist William Nicol invented the technique for producing polarized light by cutting a crystal of Iceland spar, a variety of calcite, into a special prism which became known as the Nicol prism. The addition of two such prisms to the ordinary microscope converted the instrument into a polarizing, or petrographic microscope. Using transmitted light and Nicol prisms, it was possible to determine the internal crystallographic character of very tiny mineral grains, greatly advancing the knowledge of a rock's constituents.

During the 1840s, a development by Henry C. Sorby and others firmly laid the foundation of petrography. This was a technique to study very thin slices of rock. A slice of rock was affixed to a microscope slide and then ground so thin that light could be transmitted through mineral grains that otherwise appeared opaque. The position of adjoining grains was not disturbed, thus permitting analysis of rock texture. Thin section petrography became the standard method of rock study. Since textural details contribute greatly to knowledge of the sequence of crystallization of the various mineral constituents in a rock, petrography progressed into petrogenesis and ultimately into petrology.

It was in Europe, principally in Germany, that petrography advanced in the last half of the nineteenth century.

Methods of Investigation

Macroscopic Characters

The macroscopic characters of rocks, those visible in hand-specimens without the aid of the microscope, are very varied and difficult to describe accurately and fully. The geologist in the field depends principally on them and on a few rough chemical and physical tests; and to the practical engineer, architect and quarry-master they are all-important. Although frequently insufficient in themselves to determine the true nature of a rock, they usually serve for a preliminary classification, and often give all the information needed.

With a small bottle of acid to test for carbonate of lime, a knife to ascertain the hardness of rocks and minerals, and a pocket lens to magnify their structure, the field geologist is rarely at a loss to what group a rock belongs. The fine grained species are often in-determinable in this way, and the minute mineral components of all rocks can usually be ascertained only by microscopic examination. But it is easy to see that a sandstone or grit consists of more or less rounded, water-worn sand grains and if it contains dull, weathered particles of feldspar, shining scales of mica or small crystals of calcite these also rarely escape observation. Shales and clay rocks generally are soft, fine grained, often laminated and not infrequently contain minute organisms or fragments of plants. Limestones are easily marked with a knife-blade, effervesce readily with weak cold acid and often contain entire or broken shells or other fossils. The crystalline nature of a granite or basalt is obvious at a glance, and while the former contains white or pink feldspar, clear vitreous quartz and glancing flakes of mica, the other shows yellow-green olivine, black augite, and gray stratiated plagioclase.

Other simple tools include the blowpipe (to test the fusibility of detached crystals), the goniometer, the magnet, the magnifying glass and the specific gravity balance.

Microscopic Characteristics

When dealing with unfamiliar types or with rocks so fine grained that their component minerals cannot be determined with the aid of a hand lens, a microscope is used. Characteristics observed under the microscope include colour, colour variation under plane polarised light (pleochroism, produced by the lower Nicol prism, or more recently polarising films), fracture characteristics of the grains, refractive index (in comparison to the mounting adhesive, typically Canada Balsam), and optical symmetry (birefringent or isotropic). *In toto*, these characteristics are sufficient to identify the mineral, and often to quite tightly estimate its major element composition. The process of identifying minerals under the microscope is fairly subtle, but also mechanistic - it would be possible to develop an identification key that would allow a computer to do it. The more difficult and skilful part of optical petrography is identifying the interrelationships between grains and relating them to features seen in hand specimen, at outcrop, or in mapping.

Separation of Components

Separation of the ingredients of a crushed rock powder to obtain pure samples for analysis is a common approach. It may be performed with a powerful, adjustable-strength electromagnet. A weak magnetic field attracts magnetite, then haematite and other iron ores. Silicates that contain iron follow in definite order—biotite, enstatite, augite, hornblende, garnet, and similar ferro-magnesian minerals are successively abstract-ed. Finally, only the colorless, non-magnetic compounds, such as muscovite, calcite, quartz, and feldspar remain. Chemical methods also are useful.

A weak acid dissolves calcite from crushed limestone, leaving only dolomite, silicates, or quartz. Hydrofluoric acid attacks feldspar before quartz and, if used cautiously, dis-solves these and any glassy material in a rock powder before it dissolves augite or hy-persthene.

Methods of separation by specific gravity have a still wider application. The simplest of these is levigation—treatment by a current of water. Levigation is extensively em-ployed in mechanical analysis of soils and treatment of ores, but is not so successful with rocks, as their components do not, as a rule, differ greatly in specific gravity. Flu-ids are used that do not attack most rock-forming minerals, but have a high specific gravity. Solutions of potassium mercuric iodide (sp. gr. 3.196), cadmium borotungstate (sp. gr. 3.30), methylene iodide (sp. gr. 3.32), bromoform (sp. gr. 2.86), or acetylene bromide (sp. gr. 3.00) are the principal fluids employed. They may be diluted (with water, benzene, etc.) or concentrated by evaporation.

If the rock is granite consisting of biotite (sp. gr. 3.1), muscovite (sp. gr. 2.85), quartz (sp. gr. 2.65), oligoclase (sp. gr. 2.64), and orthoclase (sp. gr. 2.56), the crushed miner-als float in methylene iodide. On gradual dilution with benzene they precipitate in the order above. Simple in theory, these methods are tedious in practice, especially as it is common for one rock-making mineral to enclose another. However, expert handling of fresh and suitable rocks yields excellent results.

Chemical Analysis

In addition to naked-eye and microscopic investigation, chemical research methods are of great practical importance to the petrographer. Crushed and separated powders, ob-tained by the processes above, may be analyzed to determine chemical composition of minerals in the rock qualitatively or quantitatively. Chemical testing, and microscopic examination of minute grains is an elegant and valuable means of discriminating be-tween mineral components of fine-grained rocks.

Thus, the presence of apatite in rock-sections is established by covering a bare rock-sec-tion with ammonium molybdate solution. A turbid yellow precipitate forms over the crystals of the mineral in question (indicating the presence of phosphates). Many sili-cates are insoluble in acids and cannot be tested in this way, but others are partly dis-solved, leaving a film of gelatinous silica that can be stained with coloring matters, such as the aniline dyes (nepheline, analcite, zeolites, etc.).

Complete chemical analysis of rocks are also widely used and important, especially in describing new species. Rock analysis has of late years (largely under the influ-ence of the chemical laboratory of the United States Geological Survey) reached a high pitch of refinement and complexity. As many as twenty or twenty-five components may be determined, but for practical purposes a knowledge of the relative proportions of silica, alumina, ferrous and ferric oxides, magnesia, lime, potash, soda and water carry us a long way in determining a rock's position in the conventional classifications.

A chemical analysis is usually sufficient to indicate whether a rock is igneous or sedimentary, and in either case to accurately show what subdivision of these classes it belongs to. In the case of metamorphic rocks it often establishes whether the original mass was a sediment or of volcanic origin.

Specific Gravity

Specific gravity of rocks is determined by use of a balance and pycnometer. It is greatest in rocks containing the most magnesia, iron, and heavy metal while least in rocks rich in alkalis, silica, and water. It diminishes with weathering. Generally, the specific gravity of rocks with the same chemical composition is higher if highly crystalline and lower if wholly or partly vitreous. The specific gravity of the more common rocks range from about 2.5 to 3.2.

Archaeological Applications

Archaeologists use petrography to identify mineral components in pottery. This information ties the artifacts to geological areas where the raw materials for the pottery were obtained. In addition to clay, potters often used rock fragments, usually called "temper" or "aplastics", to modify the clay's properties. The geological information obtained from the pottery components provides insight into how potters selected and used local and non-local resources. Archaeologists are able to determine whether pottery found in a particular location was locally produced or traded from elsewhere. This kind of information, along with other evidence, can support conclusions about settlement patterns, group and individual mobility, social contacts, and trade networks. In addition, an understanding of how certain minerals are altered at specific temperatures can allow archaeological petrographers to infer aspects of the ceramic production process itself, such as minimum and maximum temperatures reached during the original firing of the pot.

Mineralogy

Mineralogy is a subject of geology specializing in the scientific study of chemistry, crystal structure, and physical (including optical) properties of minerals. Specific studies within mineralogy include the processes of mineral origin and formation, classification of minerals, their geographical distribution, as well as their utilization.

Mineralogy is a mixture of chemistry, materials science, physics and geology.

History

Early writing on mineralogy, especially on gemstones, comes from ancient Babylonia, the ancient Greco-Roman world, ancient and medieval China, and Sanskrit texts from ancient India and the ancient Islamic World. Books on the subject included the *Naturalis Historia* of Pliny the Elder, which not only described many different minerals but also explained many of their properties, and Kitab al Jawahir (Book of Precious Stones) by Persian scientist Al Biruni. The German Renaissance specialist Georgius Agricola wrote works such as *De re metallica* (*On Metals*, 1556) and *De Natura Fossilium* (*On the Nature of Rocks*, 1546) which began the scientific approach to the subject. Systematic scientific studies of minerals and rocks developed in post-Renaissance Europe. The modern study of mineralogy was founded on the principles of crystallography (the origins of geometric crystallography, itself, can be traced back to the mineralogy practiced in the eighteenth and nineteenth centuries) and to the microscopic study of rock sections with the invention of the microscope in the 17th century.

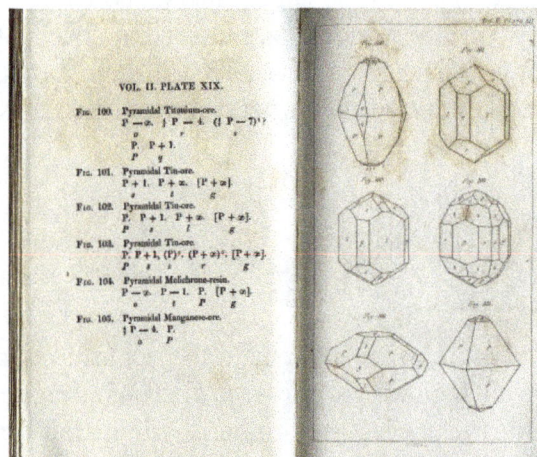

Page from *Treatise on mineralogy* by Friedrich Mohs (1825).

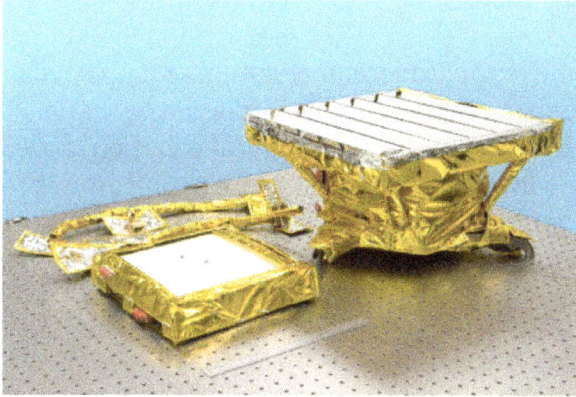

Left side of the Moon Mineralogy Mapper, a spectrometer that mapped the lunar surface.

Nicholas Steno first observed the law of constancy of interfacial angles (also known as the first law of crystallography) in quartz crystals in 1669. This was later generalized and established experimentally by Jean-Baptiste L. Romé de l'Islee in 1783. René Just Haüy, the "father of modern crystallography", showed that crystals are periodic and established that the orientations of crystal faces can be expressed in terms of rational numbers, as later encoded in the Miller indices. In 1814, Jöns Jacob Berzelius introduced a classification of minerals based on their chemistry rather than their crystal structure. William Nicol developed the Nicol prism, which polarizes light, in 1827–1828 while studying fossilized wood; Henry Clifton Sorby showed that thin sections of minerals could be identified by their optical properties using a polarizing microscope. James D. Dana published his first edition of *A System of Mineralogy* in 1837, and in a later edition introduced a chemical classification that is still the standard. X-ray diffraction was demonstrated by Max von Laue in 1912, and developed into a tool for analyzing the crystal structure of minerals by the father/son team of William Henry Bragg and William Lawrence Bragg.

More recently, driven by advances in experimental technique (such as neutron diffraction) and available computational power, the latter of which has enabled extremely accurate atomic-scale simulations of the behaviour of crystals, the science has branched out to consider more general problems in the fields of inorganic chemistry and solid-state physics. It, however, retains a focus on the crystal structures commonly encountered in rock-forming minerals (such as the perovskites, clay minerals and framework silicates). In particular, the field has made great advances in the understanding of the relationship between the atomic-scale structure of minerals and their function; in nature, prominent examples would be accurate measurement and prediction of the elastic properties of minerals, which has led to new insight into seismological behaviour of rocks and depth-related discontinuities in seismograms of the Earth's mantle. To this end, in their focus on the connection between atomic-scale phenomena and macroscopic properties, the *mineral sciences* (as they are now commonly known) display perhaps more of an overlap with materials science than any other discipline.

Physical Properties

An initial step in identifying a mineral is to examine its physical properties, many of which can be measured on a hand sample. These can be classified into density (often given as specific gravity); measures of mechanical cohesion (hardness, tenacity, cleavage, fracture, parting); macroscopic visual properties (luster, color, streak, luminescence, diaphaneity); magnetic and electric properties; radioactivity and solubility in hydrogen chloride (HCl).

If the mineral is well crystallized, it will also have a distinctive crystal habit (for example, hexagonal, columnar, botryoidal) that reflects the crystal structure or internal arrangement of atoms. It is also affected by crystal defects and twinning. Many crystals are polymorphic, having more than one possible crystal structure depending on factors such as pressure and temperature.

Crystal Structure

The crystal structure is the arrangement of atoms in a crystal. It is represented by a lattice of points which repeats a basic pattern, called a unit cell, in three dimensions. The lattice can be characterized by its symmetries and by the dimensions of the unit cell. These dimensions are represented by three *Miller indices*. The lattice remains unchanged by certain symmetry operations about any given point in the lattice: reflection, rotation, inversion, and rotary inversion, a combination of rotation and reflection. Together, they make up a mathematical object called a *crystallographic point group* or *crystal class*. There are 32 possible crystal classes. In addition, there are operations that displace all the points: translation, screw axis, and glide plane. In combination with the point symmetries, they form 230 possible space groups.

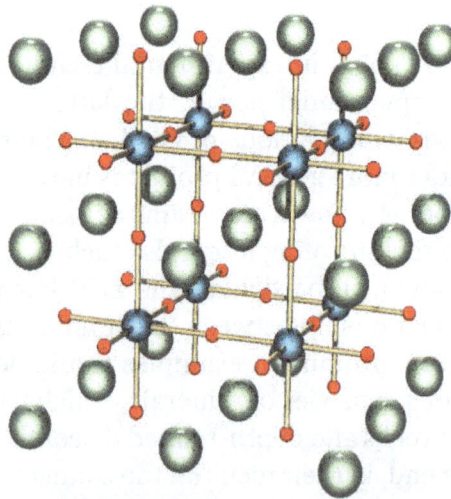

The perovskite crystal structure. The most abundant mineral in the Earth, bridgmanite, has this structure. Its chemical formula is $(Mg,Fe)SiO_3$; the red spheres are oxygen, the blue spheres silicon and the green spheres magnesium or iron.

Most geology departments have X-ray powder diffraction equipment to analyze the crystal structures of minerals. X-rays have wavelengths that are the same order of magnitude as the distances between atoms. Diffraction, the constructive and destructive interference between waves scattered at different atoms, leads to distinctive patterns of high and low intensity that depend on the geometry of the crystal. In a sample that is ground to a powder, the X-rays sample a random distribution of all crystal orientations. Powder diffraction can distinguish between minerals that may appear the same in a hand sample, for example quartz and its polymorphs tridymite and cristobalite.:54

isomorphous minerals of different compositions have similar powder diffraction patterns, the main difference being in spacing and intensity of lines. For example, the NaCl (halite) crystal structure is space group $Fm3m$; this structure is shared by sylvite (KCl), periclase (MgO), bunsenite (NiO), galena (PbS), alabandite (MnS), chlorargyrite (AgCl), and osbornite (TiN).

Chemical

A few minerals are chemical elements, including sulfur, copper, silver, and gold, but the vast majority are compounds. Before about 1947, the main method for identifying composition was *wet chemical analysis*, which involved dissolving a mineral in an acid such as hydrochloric acid (HCl). The elements in solution were then identified using colorimetry, volumetric analysis or gravimetric analysis. A variation on the wet methods is atomic absorption spectroscopy, which also requires the dissolution of the sample but is much faster and cheaper than the above methods. The solution is vaporized and its absorption spectrum is measured in the visible and ultraviolet range. Other techniques are X-ray fluorescence, electron microprobe analysis and optical emission spectrography.

Optical

In addition to macroscopic properties such as color or lustre, minerals have properties that require a polarizing microscope to observe.

Photomicrograph of olivine adcumulate, Archaean Komatiite, Agnew, Western Australia.

Transmitted Light

When light passes from air or a vacuum into a transparent crystal, some of it is reflected at the surface and some refracted. The latter is a bending of the light path that occurs because the speed of light changes as it goes into the crystal; Snell's law relates the bending angle to the Refractive index, the ratio of speed in a vacuum to speed in the crystal. Crystals whose point symmetry group falls in the cubic system are *isotropic*: the index does not depend on direction. All other crystals are *anisotropic*: light passing through them is broken up into two plane polarized rays that travel at different speeds and refract at different angles.

A polarizing microscope is similar to an ordinary microscope, but it has two plane-polarized filters, a (*polarizer*) below the sample and an analyzer above it, polarized perpendicular to each other. Light passes successively through the polarizer, the sample and the analyzer. If there is no sample, the analyzer blocks all the light from the polarizer. However, an anisotropic sample will generally change the polarization so some of the light can pass through. Thin sections and powders can be used as samples.

When an isotropic crystal is viewed, it appears dark because it does not change the polarization of the light. However, when it is immersed in a calibrated liquid with a lower index of refraction and the microscope is thrown out of focus, a bright line called a *Becke line* appears around the perimeter of the crystal. By observing the presence or absence of such lines in liquids with different indices, the index of the crystal can be estimated, usually to within ± 0.003.

Systematic

Hanksite, $Na_{22}K(SO_4)_9(CO_3)_2Cl$, one of the few minerals that is considered a carbonate and a sulfate

Systematic mineralogy is the identification and classification of minerals by their properties. Historically, mineralogy was heavily concerned with taxonomy of the rock-forming minerals. In 1959, the International Mineralogical Association formed the Commission of New Minerals and Mineral Names to rationalize the nomenclature and regulate the introduction of new names. In July 2006, it was merged with the Commission on Classification of Minerals to form the Commission on New Minerals, Nomenclature, and Classification. There are over 6,000 named and unnamed minerals, and about 100 are discovered each year. The *Manual of Mineralogy* places minerals in the following classes: native elements, sulfides, sulfosalts, oxides and hydroxides, halides, carbonates, nitrates and borates, sulfates, chromates, molybdates and tungstates, phosphates, arsenates and vanadates, and silicates.

Formation Environments

The environments of mineral formation and growth are highly varied, ranging from slow crystallization at the high temperatures and pressures of igneous melts deep within the Earth's crust to the low temperature precipitation from a saline brine at the Earth's surface.

Various possible methods of formation include:

- sublimation from volcanic gases

- deposition from aqueous solutions and hydrothermal brines

- crystallization from an igneous magma or lava

- recrystallization due to metamorphic processes and metasomatism

- crystallization during diagenesis of sediments

- formation by oxidation and weathering of rocks exposed to the atmosphere or within the soil environment.

Biomineralogy

Biomineralogy is a cross-over field between mineralogy, paleontology and biology. It is the study of how plants and animals stabilize minerals under biological control, and the sequencing of mineral replacement of those minerals after deposition. It uses techniques from chemical mineralogy, especially isotopic studies, to determine such things as growth forms in living plants and animals as well as things like the original mineral content of fossils.

A new approach to mineralogy called "mineral evolution" explores the co-evolution of the geosphere and biosphere, including the role of minerals in the origin of life and processes as mineral-catalyzed organic synthesis and the selective adsorption of organic molecules on mineral surfaces.

Uses

Minerals are essential to various needs within human society, such as minerals used as ores for essential components of metal products used in various commodities and machinery, essential components to building materials such as limestone, marble, granite, gravel, glass, plaster, cement, etc. Minerals are also used in fertilizers to enrich the growth of agricultural crops.

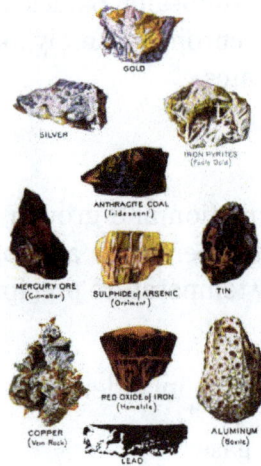

A color chart of some raw forms of commercially valuable metals.

Collecting

Mineral collecting is also a recreational study and collection hobby, with clubs and societies representing the field. Museums, such as the Smithsonian National Museum of Natural History Hall of Geology, Gems, and Minerals, the Natural History Museum of Los Angeles County, the Natural History Museum, London, and the private Mim Mineral Museum in Beirut, Lebanon, have popular collections of mineral specimens on permanent display.

Structural Geology

Structural geology is the study of the three-dimensional distribution of rock units with respect to their deformational histories. The primary goal of structural geology is to use measurements of present-day rock geometries to uncover information about the history of deformation (strain) in the rocks, and ultimately, to understand the stress field that resulted in the observed strain and geometries. This understanding of the dynamics of the stress field can be linked to important events in the geologic past; a common goal is to understand the structural evolution of a particular area with respect to regionally widespread patterns of rock deformation (e.g., mountain building, rifting) due to plate tectonics.

Use and Importance

The study of geologic structures has been of prime importance in economic geology, both petroleum geology and mining geology. Folded and faulted rock strata commonly form traps that accumulate and concentrate fluids such as petroleum and natural gas. Similarly, faulted and structurally complex areas are notable as permeable zones for hydrothermal fluids, resulting in concentrated areas of base and precious metal ore deposits. Veins of minerals containing various metals commonly occupy faults and fractures in structurally complex areas. These structurally fractured and faulted zones often occur in association with intrusive igneous rocks. They often also occur around geologic reef complexes and collapse features such as ancient sinkholes. Deposits of gold, silver, copper, lead, zinc, and other metals, are commonly located in structurally complex areas.

Structural geology is a critical part of engineering geology, which is concerned with the physical and mechanical properties of natural rocks. Structural fabrics and defects such as faults, folds, foliations and joints are internal weaknesses of rocks which may affect the stability of human engineered structures such as dams, road cuts, open pit mines and underground mines or road tunnels.

Geotechnical risk, including earthquake risk can only be investigated by inspecting a combination of structural geology and geomorphology. In addition, areas of karst landscapes which reside atop underground caverns, potential sinkholes, or other collapse features are of particular importance for these scientists. In addition, areas of steep slopes are potential collapse or landslide hazards.

Environmental geologists and hydrogeologists (also referred to as hydrologists) need to apply the tenets of structural geology to understand how geologic sites impact (or are impacted by) groundwater flow and penetration. For instance, a hydrologist may need to determine if seepage of toxic substances from waste dumps is occurring in a residential area or if salty water is seeping into an aquifer.

Plate tectonics is a theory developed during the 1960s which describes the movement of continents by way of the separation and collision of crustal plates. It is in a sense structural geology on a planet scale, and is used throughout structural geology as a framework to analyze and understand global, regional, and local scale features.

Methods

Structural geologists use a variety of methods to (first) measure rock geometries, (second) reconstruct their deformational histories, and (third) calculate the stress field that resulted in that deformation.

Geometries

Primary data sets for structural geology are collected in the field. Structural geologists measure a variety of planar features (bedding planes, foliation planes, fold axial planes, fault planes, and joints), and linear features (stretching lineations, in which minerals are ductilely extended; fold axes; and intersection lineations, the trace of a planar feature on another planar surface).

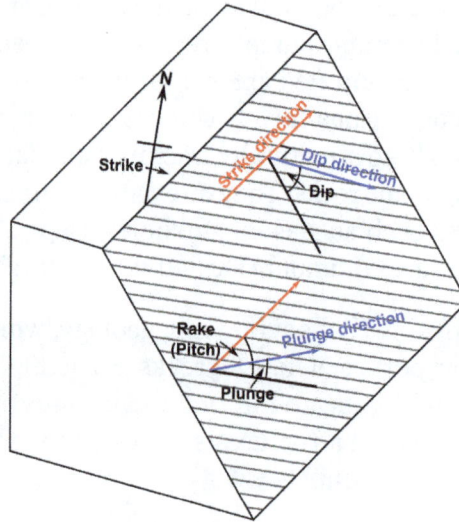

Illustration of measurement conventions for planar and linear structures

Measurement Conventions

The inclination of a planar structure in geology is measured by *strike and dip*. The strike is the line of intersection between the planar feature and a horizontal plane, taken according to the right hand convention, and the dip is the magnitude of the inclination, below horizontal, at right angles to strike. For example; striking 25 degrees East of North, dipping 45 degrees Southeast, recorded as N25E,45SE. Alternatively, dip and dip direction may be used as this is absolute. Dip direction is measured in 360 degrees, generally clockwise from North. For example, a dip of 45 degrees towards 115 degrees azimuth, recorded as 45/115. Note that this is the same as above.

The term *hade* is occasionally used and is the deviation of a plane from vertical i.e. ($90°$-dip).

Fold axis plunge is measured in dip and dip direction (strictly, plunge and azimuth of plunge). The orientation of a fold axial plane is measured in strike and dip or dip and dip direction.

Lineations are measured in terms of dip and dip direction, if possible. Often lineations occur expressed on a planar surface and can be difficult to measure directly. In this case, the lineation may be measured from the horizontal as a *rake* or *pitch* upon the surface.

Rake is measured by placing a protractor flat on the planar surface, with the flat edge horizontal and measuring the angle of the lineation clockwise from horizontal. The orientation of the lineation can then be calculated from the rake and strike-dip information of the plane it was measured from, using a stereographic projection.

If a fault has lineations formed by movement on the plane, e.g.; slickensides, this is recorded as a lineation, with a rake, and annotated as to the indication of throw on the fault.

Generally it is easier to record strike and dip information of planar structures in dip/dip direction format as this will match all the other structural information you may be recording about folds, lineations, etc., although there is an advantage to using different formats that discriminate between planar and linear data.

Plane, Fabric, Fold and Deformation Conventions

The convention for analysing structural geology is to identify the planar structures, often called *planar fabrics* because this implies a textural formation, the linear structures and, from analysis of these, unravel deformations.

Planar structures are named according to their order of formation, with original sedimentary layering the lowest at So. Often it is impossible to identify So in highly deformed rocks, so numbering may be started at an arbitrary number or given a letter (S_A, for instance). In cases where there is a bedding-plane foliation caused by burial metamorphism or diagenesis this may be enumerated as Soa.

If there are folds, these are numbered as F_1, F_2, etc. Generally the axial plane foliation or cleavage of a fold is created during folding, and the number convention should match. For example, an F_2 fold should have an S_2 axial foliation.

Deformations are numbered according to their order of formation with the letter D denoting a deformation event. For example, D_1, D_2, D_3. Folds and foliations, because they are formed by deformation events, should correlate with these events. For example, an F_2 fold, with an S_2 axial plane foliation would be the result of a D_2 deformation.

Metamorphic events may span multiple deformations. Sometimes it is useful to identify them similarly to the structural features for which they are responsible, e.g.; M_2. This may be possible by observing porphyroblast formation in cleavages of known deformation age, by identifying metamorphic mineral assemblages created by different events, or via geochronology.

Intersection lineations in rocks, as they are the product of the intersection of two planar structures, are named according to the two planar structures from which they are formed. For instance, the intersection lineation of a S_1 cleavage and bedding is the L_{1-0} intersection lineation (also known as the cleavage-bedding lineation).

Stretching lineations may be difficult to quantify, especially in highly stretched ductile rocks where minimal foliation information is preserved. Where possible, when correlated with deformations (as few are formed in folds, and many are not strictly associated with planar foliations), they may be identified similar to planar surfaces and folds, e.g.; L_1, L_2. For convenience some geologists prefer to annotate them with a subscript S, for example L_{s1} to differentiate them from intersection lineations, though this is generally redundant.

Stereographic Projections

Stereographic projection of structural strike and dip measurements is a powerful method for analyzing the nature and orientation of deformation stresses, lithological units and penetrative fabrics.

Rock Macro-structures

On a large scale, structural geology is the study of the three-dimensional interaction and relationships of stratigraphic units within terranes of rock or geological regions.

This branch of structural geology deals mainly with the orientation, deformation and relationships of stratigraphy (bedding), which may have been faulted, folded or given a foliation by some tectonic event. This is mainly a geometric science, from which *cross sections* and three-dimensional *block models* of rocks, regions, terranes and parts of the Earth's crust can be generated.

Study of regional structure is important in understanding orogeny, plate tectonics and more specifically in the oil, gas and mineral exploration industries as structures such as faults, folds and unconformities are primary controls on ore mineralisation and oil traps.

Modern regional structure is being investigated using seismic tomography and seismic reflection in three dimensions, providing unrivaled images of the Earth's interior, its faults and the deep crust. Further information from geophysics such as gravity and airborne magnetics can provide information on the nature of rocks imaged to be in the deep crust.

Rock Microstructures

Rock microstructure or *texture* of rocks is studied by structural geologists on a small scale to provide detailed information mainly about metamorphic rocks and some features of sedimentary rocks, most often if they have been folded.

Textural study involves measurement and characterisation of foliations, crenulations, metamorphic minerals, and timing relationships between these structural features and mineralogical features.

Usually this involves collection of hand specimens, which may be cut to provide petrographic thin sections which are analysed under a petrographic microscope.

Microstructural analysis finds application also in multi-scale statistical analysis, aimed to analyze some rock features showing scale invariance.

Kinematics

Geologists use rock geometry measurements to understand the history of strain in rocks. Strain can take the form of brittle faulting and ductile folding and shearing.

Brittle deformation takes place in the shallow crust, and ductile deformation takes place in the deeper crust, where temperatures and pressures are higher.

Stress Fields

By understanding the constitutive relationships between stress and strain in rocks, geologists can translate the observed patterns of rock deformation into a stress field during the geologic past. The following list of features are typically used to determine stress fields from deformational structures.

- In perfectly brittle rocks, faulting occurs at 30° to the greatest compressional stress. (Byerlee's Law)

- The greatest compressive stress is normal to fold axial planes.

Stratigraphy

Stratigraphy is a branch of geology which studies rock layers (strata) and layering (stratification). It is primarily used in the study of sedimentary and layered volcanic rocks. Stratigraphy includes two related subfields: lithologic stratigraphy or lithostratigraphy, and biologic stratigraphy or biostratigraphy.

The Permian through Jurassic strata of the Colorado Plateau area of southeastern Utah demonstrate the principles of stratigraphy.

Historical Development

Nicholas Steno established the theoretical basis for stratigraphy when he introduced the law of superposition, the principle of original horizontality and the principle of lateral continuity in a 1669 work on the fossilization of organic remains in layers of sediment.

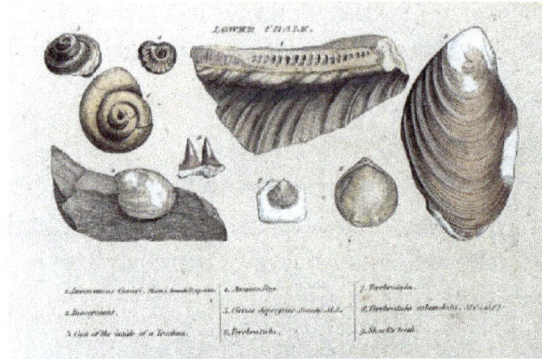

Engraving from William Smith's monograph on identifying strata based on fossils

The first practical large-scale application of stratigraphy was by William Smith in the 1790s and early 19th century. Smith, known as the "Father of English geology", created the first geologic map of England and first recognized the significance of strata or rock layering and the importance of fossil markers for correlating strata. Another influential application of stratigraphy in the early 19th century was a study by Georges Cuvier and Alexandre Brongniart of the geology of the region around Paris.

Strata in Cafayate (Argentina)

Lithostratigraphy

Lithostratigraphy, or lithologic stratigraphy, provides the most obvious visible layering. It deals with the physical contrasts in lithology, or rock type. Such layers can occur both vertically – in layering or bedding of varying rock types – and laterally – reflecting changing environments of deposition (known as facies change). Key concepts in stratigraphy involve understanding how certain geometric relationships between rock layers arise and what these geometries mean in terms of the depositional environment. Stratigraphers have codified a basic concept of their discipline in the law of superposition, which simply states that, in an undeformed stratigraphic sequence, the oldest strata occur at the base of the sequence.

Chemostratigraphy studies the changes in the relative proportions of trace elements and isotopes within and between lithologic units. Carbon and oxygen isotope ratios vary with time, and researchers can use them to map subtle changes that occurred in the paleoenvironment. This has led to the specialized field of isotopic stratigraphy.

Chalk layers in Cyprus, showing sedimentary layering

Cyclostratigraphy documents the often cyclic changes in the relative proportions of minerals (particularly carbonates), grain size, or thickness of sediment layers (varves) and of fossil diversity with time, related to seasonal or longer term changes in palaeo-climates.

Biostratigraphy

Biostratigraphy or paleontologic stratigraphy is based on fossil evidence in the rock layers. Strata from widespread locations containing the same fossil fauna and flora are correlatable in time. Biologic stratigraphy was based on William Smith's principle of faunal succession, which predated, and was one of the first and most powerful lines of evidence for, biological evolution. It provides strong evidence for the formation (speciation) and extinction of species. The geologic time scale was developed during the 19th century, based on the evidence of biologic stratigraphy and faunal succession. This timescale remained a relative scale until the development of radiometric dating, which gave it and the stratigraphy it was based on an absolute time framework, leading to the development of chronostratigraphy.

One important development is the Vail curve, which attempts to define a global historical sea-level curve according to inferences from worldwide stratigraphic patterns. Stratigraphy is also commonly used to delineate the nature and extent of hydrocarbon-bearing reservoir rocks, seals, and traps in petroleum geology.

Chronostratigraphy

Chronostratigraphy is the branch of stratigraphy that studies the absolute, not relative, age of rock strata. The branch is concerned with deriving geochronological data for rock units, both directly and inferentially, so that a sequence of time-relative events of rocks within a region can be derived. In essence, chronostratigraphy seeks to understand the geologic history of rocks and regions.

The ultimate aim of chronostratigraphy is to arrange the sequence of deposition and the time of deposition of all rocks within a geological region and, eventually, the entire geologic record of the earth.

A gap or missing strata in the geological record of an area is called a stratigraphic hiatus. This may be the result of lack of sediment deposition or it may be due to removal by erosion, in which case it may be called a vacuity. It is called a *hiatus* because deposition was *on hold* for a period of time. A physical gap may represent both a period of non-deposition and a period of erosion. A fault may cause the appearance of a hiatus.

Magnetostratigraphy

Magnetostratigraphy is a chronostratigraphic technique used to date sedimentary and volcanic sequences. The method works by collecting oriented samples at measured intervals throughout a section. The samples are analyzed to determine their detrital remanent magnetism (DRM), that is, the polarity of Earth's magnetic field at the time a stratum was deposited. For sedimentary rocks, this is possible because, when very fine-grained magnetic minerals (< 17 μm) fall through the water column, they orient themselves with Earth's magnetic field. Upon burial, that orientation is preserved. The minerals behave like tiny compasses. For volcanic rocks, magnetic minerals, which form in the melt, are fixed in place upon crystallization or freezing of the lava and are oriented with the ambient magnetic field.

Oriented paleomagnetic core samples are collected in the field; mudstones, siltstones, and very fine-grained sandstones are the preferred lithologies because the magnetic grains are finer and more likely to orient with the ambient field during deposition. If the ancient magnetic field were oriented similar to today's field (North Magnetic Pole near the North Rotational Pole), the strata would retain a normal polarity. If the data indicate that the North Magnetic Pole were near the South Rotational Pole, the strata would exhibit reversed polarity.

Results of the individual samples are analyzed by removing the natural remanent magnetization (NRM) to reveal the DRM. Following statistical analysis, the results are used to generate a local magnetostratigraphic column that can then be compared against the Global Magnetic Polarity Time Scale.

This technique is used to date sequences that generally lack fossils or interbedded igneous rocks. The continuous nature of the sampling means that it is also a powerful technique for the estimation of sediment-accumulation rates.

Archaeological Stratigraphy

In the field of archaeology, soil stratigraphy is used to better understand the processes that form and protect archaeological sites. Since the law of superposition holds true, it can help date finds or features from each context; these finds and features can be placed in sequence and the dates interpolated. Phases of activity can also often be seen through stratigraphy, especially when a trench or feature is viewed in section (profile). Because pits and other features can be dug down into earlier levels, not all material at the same absolute depth is necessarily of the same age; close attention has to be paid to the archeological layers. The Harris-matrix is a tool to depict complex stratigraphic relations when they are found, for example, in the context of urban archaeology.

Lithostratigraphy

Lithostratigraphy is a sub-discipline of stratigraphy, the geological science associated with the study of strata or rock layers. Major focuses include geochronology, comparative geology, and petrology. In general a stratum will be primarily igneous or sedimentary relating to how the rock was formed.

The Permian through Jurassic lithostratigraphy of the Colorado Plateau area of southeastern Utah that makes up much of the famous prominent rock formations in protected areas such as Capitol Reef National Park and Canyonlands National Park. From top to bottom: Rounded tan domes of the Navajo Sandstone, layered red Kayenta Formation, cliff-forming, vertically-jointed, red Wingate Sandstone, slope-forming, purplish Chinle Formation, layered, lighter-red Moenkopi Formation, and white, layered Cutler Formation sandstone. Picture from Glen Canyon National Recreation Area, Utah.

Strata in Salta (Argentina).

Sedimentary layers are laid down by deposition of sediment associated with weathering processes, decaying organic matters (biogenic) or through chemical precipitation. These layers are distinguishable as having many fossils and are important for the study of biostratigraphy. Igneous layers are either plutonic or volcanic in character depending upon the cooling rate of the rock. These layers are generally devoid of fossils and represent intrusions and volcanic activity that occurred over the geologic history of the area.

There are a number of principles that are used to explain the appearance of stratum. When an igneous rock cuts across a formation of sedimentary rock, then we can say that the igneous intrusion is younger than the sedimentary rock. The principle of superposition states that a sedimentary rock layer in a tectonically undisturbed stratum is younger than the one beneath and older than the one above it. The principle of original horizontality states that the deposition of sediments occurs as essentially horizontal beds.

Types of Lithostratigraphic Units

A lithostratigraphic unit conforms to the law of superposition, which state that in any succession of strata, not disturbed or overturned since deposition, younger rocks lies above older rocks. The principle of lateral continuity states that a set of bed extends and can be traceable over a large area.

Lithostratigraphic units are recognized and defined on the basis of observable rock characteristics. The descriptions of strata based on physical appearance define facies. Lithostratigraphic units are only defined by lithic characteristics, and not by age.

Stratotype: A designated type of unit consisting of accessible rocks that contain clear-cut characteristics which are representative of a particular lithostratigraphic unit.

Lithosome: Masses of rock of essentially uniform character and having interchanging relationships with adjacent masses of different lithology. e.g.: shale lithosome, limestone lithosome.

The fundamental Lithostratigraphic unit is the formation. A formation is a lithologically distinctive stratigraphic unit that is large enough to be mappable and traceable. Formations may be subdivided into members and beds and aggregated with other formations into groups and supergroups.

Stratigraphic Relationship

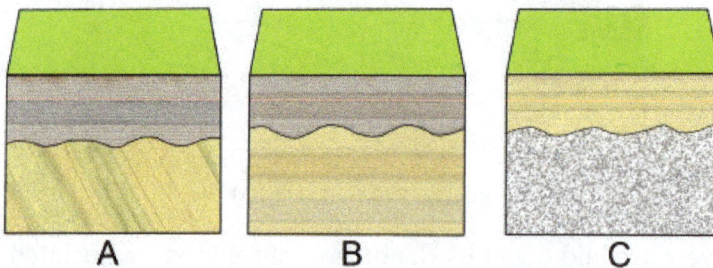

Diagrams showing stratigraphic relations: A: an angular unconformity;
B: a disconformity; C: a nonconformity.

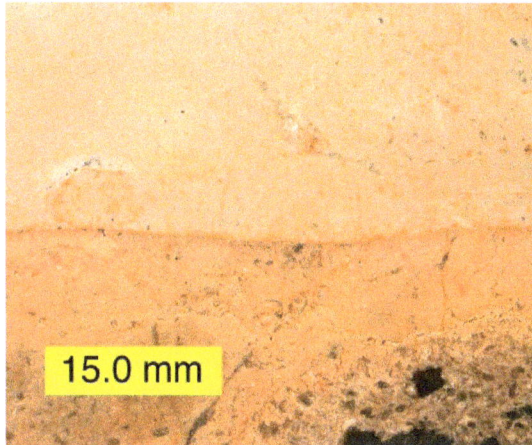

Disconformity with the Lower Cretaceous Edwards Formation overlying a Lower Permian limestone; hiatus is about 165 million years; Texas.

Two types of contact: *conformable* and *unconformable*.

Conformable: unbroken deposition, no break or hiatus (break or interruption in the continuity of the geological record). The surface strata resulting is called a *conformity*.

Two types of contact between conformable strata: *abrupt contacts* (directly separate beds of distinctly different lithology, minor depositional break, called *diastems*) and *gradational contact* (gradual change in deposition, mixing zone).

Unconformable: period of erosion/non-deposition. The surface stratum resulting is called an unconformity.

Four types of unconformity:

- *Angular unconformity*: younger sediment lies upon an eroded surface of tilted or folded older rocks. The older rock dips at a different angle from the younger.

- *Disconformity*: the contact between younger and older beds is marked by visible, irregular erosional surfaces. Paleosol might develop right above the disconformity surface because of the non-deposition setting.

- *Paraconformity*: the bedding planes below and above the unconformity are parallel. A time gap is present, as shown by a faunal break, but there is no erosion, just a period of non-deposition.

- *Nonconformity*: relatively young sediments are deposited right above older igneous or metamorphic rocks.

Lithostratigraphic Correlation

To correlate lithostratigraphic units, geologists define facies, and look for key beds or key sequences that can be used as a datum.

- Direct correlation: based on lithology, color, structure, thickness...

- Indirect correlation: electric log correlation (gamma-ray, density, resistivity...)

Biostratigraphy

Biostratigraphy is the branch of stratigraphy which focuses on correlating and assigning relative ages of rock strata by using the fossil assemblages contained within them. Usually the aim is correlation, demonstrating that a particular horizon in one geological section represents the same period of time as another horizon at some other section. The fossils are useful because sediments of the same age can look completely different because of local variations in the sedimentary environment. For example, one section might have been made up of clays and marls while another has more chalky limestones, but if the fossil species recorded are similar, the two sediments are likely to have been laid down at the same time.

Biostratigraphy originated in the early 19th century, where geologists recognised that the correlation of fossil assemblages between rocks of similar type but different age decreased as the difference in age increased. The method was well-established before Charles Darwin explained the mechanism behind it - evolution.

The first reef builder is a worldwide index fossil for the Lower Cambrian

Ammonites, graptolites, archeocyathids, and trilobites are index fossils that are widely used in biostratigraphy. Microfossils such as acritarchs, chitinozoans, conodonts, dinoflagellate cysts, pollen, spores and foraminiferans are also frequently used. Different fossils work well for sediments of different ages; trilobites, for example, are particularly useful for sediments of Cambrian age. To work well, the fossils used must be widespread geographically, so that they can occur in many different places. They must also be short lived as a species, so that the period of time during which they could be incorporated in the sediment is relatively narrow. The longer lived the species, the poorer the stratigraphic precision, so fossils that evolve rapidly, such as ammonites, are favoured over forms that evolve much more slowly, like nautiloids. Often biostratigraphic correlations are based on a fauna, not an individual species, as this allows greater precision. Further, if only one species is present in a sample, it can mean that (1) the strata were formed in the known fossil range of that organism; (2) that the fossil range of the organism was incompletely known, and the strata extend the known fossil range. For instance, the presence of the fossil *Treptichnus pedum* was used to define the base of the Cambrian period, but it has since been found in older strata.

Fossil assemblages were traditionally used to designate the duration of periods. Since a large change in fauna was required to make early stratigraphers create a new period, most of the periods we recognise today are terminated by a major extinction event or faunal turnover.

Fossils as a Basis for Stratigraphic Subdivision

Concept of Stage

A stage is a major subdivision of strata, each systematically following the other each bearing a unique assemblage of fossils. Therefore, stages can be defined as a group of strata containing the same major fossil assemblages. French palaeontologist Alcide d'Orbigny is credited for the invention of this concept. He named stages after geographic localities with particularly good sections of rock strata that bear the characteristic fossils on which the stages are based.

Concept of Zone

In 1856 German palaeontologist Albert Oppel introduced the concept of zone (also known as biozones or Oppel zone). A zone includes strata characterised by the over-lapping range of fossils. They represent the time between the appearance of species chosen at the base of the zone and the appearance of other species chosen at the base of the next succeeding zone. Oppel's zones are named after a particular distinctive fossil species, called an index fossil. Index fossils are one of the species from the assemblage of species that characterise the zone.

The zone is the fundamental biostratigraphic unit. Its thickness range from a few to hundreds of metres, and its extant range from local to worldwide. Biostratigraphic units are divided into six principal kinds of biozones:

- *Taxon range biozone* represent the known stratigraphic and geographic range of occurrence of a single taxon.

- *Concurrent range biozone* include the concurrent, coincident, or overlapping part of the range of two specified taxa.

- *Interval biozone* include the strata between two specific biostratigraphic surfaces. It can be based on lowest or highest occurrences.

- *Lineage biozone* are strata containing species representing a specific segment of an evolutionary lineage.

- *Assemblage biozones* are strata that contain a unique association of three or more taxa.

- *Abundance biozone* are strata in which the abundance of a particular taxon or group of taxa is significantly greater than in the adjacent part of the section.

Index Fossils

Amplexograptus, a graptolite index fossil, from the Ordovician near Caney Springs, Tennessee.

To be useful in stratigraphic correlation index fossils should be:

- Independent of their environment

- Geographically widespread (provincialism/isolation of species should be avoided as much as possible)

- Rapidly evolving

- Abundant (easy to find in the rock record)

- Easy to preserve (Easier in low-energy, non-oxidized environment)

- Easy to identify

Chronostratigraphy

Chronostratigraphy is the branch of stratigraphy that studies the age of rock strata in relation to time.

The ultimate aim of chronostratigraphy is to arrange the sequence of deposition and the time of deposition of all rocks within a geological region, and eventually, the entire geologic record of the Earth.

The standard stratigraphic nomenclature is a chronostratigraphic system based on palaeontological intervals of time defined by recognised fossil assemblages (biostratigraphy). The aim of chronostratigraphy is to give a meaningful age date to these fossil assemblage intervals and interfaces.

Methodology

Chronostratigraphy relies heavily upon isotope geology and geochronology to derive hard dating of known and well defined rock units which contain the specific fossil assemblages defined by the stratigraphic system. As it is practically very difficult to isotopically date most fossils and sedimentary rocks directly, inferences must be made in order to arrive at an age date which reflects the beginning of the interval.

The methodology used is derived from the law of superposition and the principles of cross-cutting relationships.

Because igneous rocks occur at specific intervals in time and are essentially instantaneous on a geologic time scale, and because they contain mineral assemblages which may be dated more accurately and precisely by isotopic methods, the construction of a chronostratigraphic column relies heavily upon intrusive and extrusive igneous rocks.

Metamorphism, often associated with faulting, may also be used to bracket depositional intervals in a chronostratigraphic column. Metamorphic rocks can occasionally be dated, and this may give some limits to the age at which a bed could have been laid down. For example, if a bed containing graptolites overlies crystalline basement at some point, dating the crystalline basement will give a maximum age of that fossil assemblage.

This process requires a considerable degree of effort and checking of field relationships and age dates. For instance, there may be many millions of years between a bed being laid down and an intrusive rock cutting it; the estimate of age must necessarily be between the oldest cross-cutting intrusive rock in the fossil assemblage and the youngest rock upon which the fossil assemblage rests.

Units

Chronostratigraphic units, with examples:

- eonothem – Phanerozoic

- erathem – Paleozoic

- system – Ordovician

- series – Upper Ordovician

- stage – Ashgill

Differences between Chronostratigraphy and Geochronology

It is important not to confuse geochronologic and chronostratigraphic units. Chronostratigraphic units are geological material, so it is correct to say that fossils of the species

Tyrannosaurus rex have been found in the Upper Cretaceous Series. Geochronological units are periods of time and take the same name as standard stratigraphic units but replacing the terms upper/lower with late/early. Thus it is also correct to say that *Tyrannosaurus rex* lived during the Late Cretaceous Epoch.

Chronostratigraphy is an important branch of stratigraphy because the age correlations derived are crucial to drawing accurate cross sections of the spatial organization of rocks and to preparing accurate paleogeographic reconstructions.

Magnetostratigraphy

Magnetostratigraphy is a geophysical correlation technique used to date sedimentary and volcanic sequences. The method works by collecting oriented samples at measured intervals throughout the section. The samples are analyzed to determine their *characteristic remanent magnetization* (ChRM), that is, the polarity of Earth's magnetic field at the time a stratum was deposited. This is possible because volcanic flows acquire a thermoremanent magnetization and sediments acquire a depositional remanent magnetization, both of which reflect the direction of the Earth's field at the time of formation.

Technique

When measurable magnetic properties of rocks vary stratigraphically they may be the basis for related but different kinds of stratigraphic units known collectively as *magnetostratigraphic units (magnetozones)*. The magnetic property most useful in stratigraphic work is the change in the direction of the remanent magnetization of the rocks, caused by reversals in the polarity of the Earth's magnetic field. The direction of the remnant magnetic polarity recorded in the stratigraphic sequence can be used as the basis for the subdivision of the sequence into units characterized by their magnetic polarity. Such units are called "magnetostratigraphic polarity units" or chrons.

If the ancient magnetic field was oriented similar to today's field (North Magnetic Pole near the Geographic North Pole) the strata retain a normal polarity. If the data indicate that the North Magnetic Pole was near the Geographic South Pole, the strata exhibit reversed polarity.

Sampling Procedures

Oriented paleomagnetic samples are collected in the field using a rock core drill, or as *hand samples* (chunks broken off the rock face). To average out sampling errors, a minimum of three samples is taken from each sample site. Spacing of the sample sites within a stratigraphic section depends on the rate of deposition and the age of the section. In sedimentary layers, the preferred lithologies are mudstones, claystones, and very fine-grained siltstones because the magnetic grains are finer and more likely to orient with the ambient field during deposition.

Analytical Procedures

Samples are first analyzed in their natural state to obtain their natural remanent magnetization (NRM). The NRM is then stripped away in a stepwise manner using thermal or alternating field demagnetization techniques to reveal the stable magnetic component.

Magnetic orientations of all samples from a site are then compared and their average magnetic polarity is determined with directional statistics, most commonly Fisher statistics or bootstrapping. The statistical significance of each average is evaluated. The latitudes of the Virtual Geomagnetic Poles from those sites determined to be statistically significant are plotted against the stratigraphic level at which they were collected. These data are then abstracted to the standard black and white magnetostratigraphic columns in which black indicates normal polarity and white is reversed polarity.

Correlation and Ages

Because the polarity of a stratum can only be normal or reversed, variations in the rate at which the sediment accumulated can cause the thickness of a given polarity zone to vary from one area to another. This presents the problem of how to correlate zones of like polarities between different stratigraphic sections. To avoid confusion at least one isotopic age needs to be collected from each section. In sediments, this is often obtained from layers of volcanic ash. Failing that, one can tie a polarity to a biostratigraphic event that has been correlated elsewhere with isotopic ages. With the aid of the independent isotopic age or ages, the local magnetostratigraphic column is correlated with the Global Magnetic Polarity Time Scale (GMPTS).

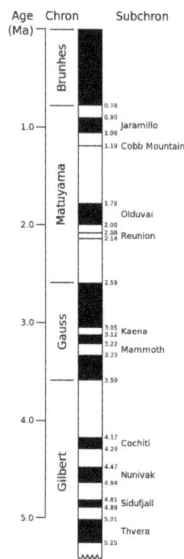

Geomagnetic polarity in late Cenozoic

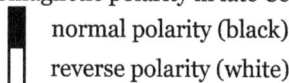

▮ normal polarity (black)
▯ reverse polarity (white)

Because the age of each reversal shown on the GMPTS is relatively well known, the correlation establishes numerous time lines through the stratigraphic section. These ages provide relatively precise dates for features in the rocks such as fossils, changes in sedimentary rock composition, changes in depositional environment, etc. They also constrain the ages of cross-cutting features such as faults, dikes, and unconformities.

Sediment Accumulation Rates

Perhaps the most powerful application of these data is to determine the rate at which the sediment accumulated. This is accomplished by plotting the age of each reversal (in millions of years ago) vs. the stratigraphic level at which the reversal is found (in meters). This provides the rate in meters per million years which is usually rewritten in terms of millimeters per year (which is the same as kilometers per million years).

These data are also used to model basin subsidence rates. Knowing the depth of a hydrocarbon source rock beneath the basin-filling strata allows calculation of the age at which the source rock passed through the generation window and hydrocarbon migration began. Because the ages of cross-cutting trapping structures can usually be determined from magnetostratigraphic data, a comparison of these ages will assist reservoir geologists in their determination of whether or not a play is likely in a given trap.

Changes in sedimentation rate revealed by magnetostratigraphy are often related to either climatic factors or to tectonic developments in nearby or distant mountain ranges. Evidence to strengthen this interpretation can often be found by looking for subtle changes in the composition of the rocks in the section. Changes in sandstone composition are often used for this type of interpretation.

References

- Blatt, Harvey; Tracy, Robert J.; Owens, Brent (2005), Petrology: igneous, sedimentary, and metamorphic (New York: W. H. Freeman). ISBN 978-0-7167-3743-8

- Dietrich, Richard Vincent; Skinner, Brian J. (2009), Gems, Granites, and Gravels: knowing and using rocks and minerals (Cambridge University Press). ISBN 978-0-521-10722-8

- Fei, Yingwei; Bertka, Constance M.; Mysen, Bjorn O. (eds.) (1999), Mantle Petrology: field observations and high-pressure experimentation (Houston TX: Geochemical Society). ISBN 0-941809-05-6

- Philpotts, Anthony; Ague, Jay (2009), Principles of Igneous and Metamorphic Petrology (Cambridge University Press). ISBN 978-0-521-88006-0

- Needham, Joseph (1959). Science and civilisation in China. Cambridge: Cambridge University Press. pp. 637–638. ISBN 978-0521058018.

- Nesse, William D. (2012). Introduction to mineralogy (2nd ed.). New York: Oxford University Press. ISBN 978-0199827381.

- Rafferty, John P. (2012). Geological sciences (1st ed.). New York: Britannica Educational Pub. in association with Rosen Educational Services. pp. 14–15. ISBN 9781615304950.

- Klein, Cornelis; Philpotts, Anthony R. (2013). Earth materials : introduction to mineralogy and petrology. New York: Cambridge University Press. ISBN 9780521145213.

- B.A. van der Pluijm and S. Marshak (2004). Earth Structure - An Introduction to Structural Geology and Tectonics (2nd ed.). New York: W. W. Norton. p. 656. ISBN 0-393-92467-X.

- Davies G.L.H. (2007). Whatever is Under the Earth the Geological Society of London 1807-2007. London: Geological Society. p. 78. ISBN 9781862392144.

- Christopherson, R. W., 2008. Geosystems: An Introduction to Physical Geography, 7th ed., New York: Pearson Prentice-Hall. ISBN 978-0-13-600598-8

3

Principles of Geology

The principles of geology discussed within this section are crosscutting relationships, uniformitarianism, principle of original horizontality, law of superposition and principal of lateral continuity. One of the basic principles of geology is uniformitarianism; it is the supposition that same laws and processes that have in the past operated the universe, presently also operate it. This chapter elucidates the crucial theories and principles of geology.

Cross-cutting Relationships

Cross-cutting relationships is a principle of geology that states that the geologic feature which cuts another is the younger of the two features. It is a relative dating technique in geology. It was first developed by Danish geological pioneer Nicholas Steno in *Dissertationis prodromus* (1669) and later formulated by James Hutton in *Theory of the Earth* (1795) and embellished upon by Charles Lyell in *Principles of Geology* (1830).

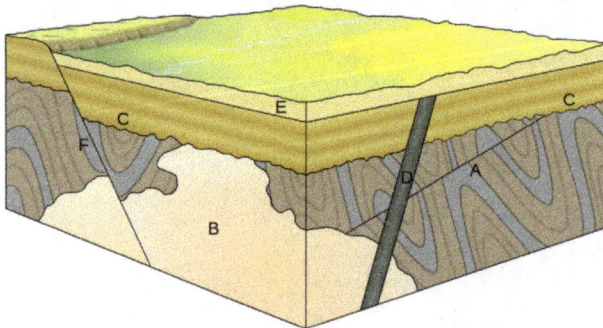

Cross-cutting relations can be used to determine the relative ages of rock strata and other geological structures. Explanations: A - folded rock strata cut by a thrust fault; B - large intrusion (cutting through A); C - erosional angular unconformity (cutting off A & B) on which rock strata were deposited; D - volcanic dike (cutting through A, B & C); E - even younger rock strata (overlying C & D); F - normal fault (cutting through A, B, C & possibly E).

Types

There are several basic types of cross cutting relationships:

- Structural relationships may be faults or fractures cutting through an older rock.

- Intrusional relationships occur when an igneous pluton or dike is intruded into pre-existing rocks.

- Stratigraphic relationships may be an erosional surface (or unconformity) cuts across older rock layers, geological structures, or other geological features.

- Sedimentological relationships occur where currents have eroded or scoured older sediment in a local area to produce, for example, a channel filled with sand.

- Paleontological relationships occur where animal activity or plant growth produces truncation. This happens, for example, where animal burrows penetrate into pre-existing sedimentary deposits.

- Geomorphological relationships may occur where a surficial feature, such as a river, flows through a gap in a ridge of rock. In a similar example, an impact crater excavates into a subsurface layer of rock.

Cross-cutting relationships may be compound in nature. For example, if a fault were truncated by an unconformity, and that unconformity cut by a dike. Based upon such compound cross-cutting relationships it can be seen that the fault is older than the unconformity which in turn is older than the dike. Using such rationale, the sequence of geological events can be better understood.

Scale

Cross-cutting relationships may be seen cartographically, megascopically, and microscopically. In other words, these relationships have various scales. A cartographic crosscutting relationship might look like, for example, a large fault dissecting the landscape on a large map. Megascopic cross-cutting relationships are features like igneous dikes, as mentioned above, which would be seen on an outcrop or in a limited geographic area. Microscopic cross-cutting relationships are those that require study by magnification or other close scrutiny. For example, penetration of a fossil shell by the drilling action of a boring organism is an example of such a relationship.

Cross-cutting relationships involving an andesitic dike in Peru that cuts across the lower sedimentary strata. Both the dike and the lower strata are cut by an unconformity

A light-gray igneous intrusion in Sweden cut by a younger white pegmatite dike, which in turn is cut by an even younger black diabase dike

Other Use

Cross-cutting relationships can also be used in conjunction with radiometric age dating to effect an age bracket for geological materials that cannot be directly dated by radiometric techniques. For example, if a layer of sediment containing a fossil of interest is bounded on the top and bottom by unconformities, where the lower unconformity truncates dike A and the upper unconformity truncates dike B (which penetrates the layer in question), this method can be used. A radiometric age date from crystals in dike A will give the maximum age date for the layer in question and likewise, crystals from dike B will give us the minimum age date. This provides an age bracket, or range of possible ages, for the layer in question.

Uniformitarianism

Hutton's Unconformity at Jedburgh. A photograph shows the current scene (2003), below John Clerk of Eldin's illustration of 1787.

Uniformitarianism is the assumption that the same natural laws and processes that operate in the universe now have always operated in the universe in the past and apply everywhere in the universe. It refers to invariance in the metaphysical principles underpinning science, such as the constancy of causal structure throughout space-time, but has also been used to describe spatiotemporal invariance of physical laws. Though an unprovable postulate that cannot be verified using the scientific method, uniformitarianism has been a key first principle of virtually all fields of science.

In geology, uniformitarianism has included the gradualistic concept that "the present is the key to the past" and is functioning at the same rates, though many modern geologists no longer hold to a strict gradualism. Coined by William Whewell, it was originally proposed in contrast to catastrophism by British naturalists in the late 18th century, starting with the work of the Scottish geologist James Hutton. Hutton's work was later refined by John Playfair and popularised by Charles Lyell's *Principles of Geology* in 1830.

History

18th Century

The earlier conceptions likely had little influence on 18th-century European geological explanations for the formation of Earth. Abraham Gottlob Werner (1749-1817) proposed Neptunism, where strata represented deposits from shrinking seas precipitated onto primordial rocks such as granite. In 1785 James Hutton proposed an opposing, self-maintaining infinite cycle based on natural history and not on the Biblical account.

Cliff at the east of Siccar Point in Berwickshire, showing the near-horizontal red sandstone layers above vertically tilted greywacke rocks.

The solid parts of the present land appear in general, to have been composed of the productions of the sea, and of other materials similar to those now found upon the shores. Hence we find reason to conclude:

1st, That the land on which we rest is not simple and original, but that it is a composition, and had been formed by the operation of second causes.

2nd, That before the present land was made, there had subsisted a world composed of sea and land, in which were tides and currents, with such operations at the bottom of the sea as now take place. And,

Lastly, That while the present land was forming at the bottom of the ocean, the former land maintained plants and animals; at least the sea was then inhabited by animals, in a similar manner as it is at present.

Hence we are led to conclude, that the greater part of our land, if not the whole had been produced by operations natural to this globe; but that in order to make this land a permanent body, resisting the operations of the waters, two things had been required;

1st, The consolidation of masses formed by collections of loose or incoherent materials;

2ndly, The elevation of those consolidated masses from the bottom of the sea, the place where they were collected, to the stations in which they now remain above the level of the ocean.

Hutton then sought evidence to support his idea that there must have been repeated cycles, each involving deposition on the seabed, uplift with tilting and erosion, and then moving undersea again for further layers to be deposited. At Glen Tilt in the Cairngorm mountains he found granite penetrating metamorphic schists, in a way which indicated to him that the presumed primordial rock had been molten after the strata had formed. He had read about angular unconformities as interpreted by Neptunists, and found an unconformity at Jedburgh where layers of greywacke in the lower layers of the cliff face have been tilted almost vertically before being eroded to form a level plane, under horizontal layers of Old Red Sandstone. In the spring of 1788 he took a boat trip along the Berwickshire coast with John Playfair and the geologist Sir James Hall, and found a dramatic unconformity showing the same sequence at Siccar Point. Playfair later recalled that "the mind seemed to grow giddy by looking so far into the abyss of time", and Hutton concluded a 1788 paper he presented at the Royal Society of Edinburgh, later rewritten as a book, with the phrase "we find no vestige of a beginning, no prospect of an end".

Both Playfair and Hall wrote their own books on the theory, and for decades robust debate continued between Hutton's supporters and the Neptunists. Georges Cuvier's paleontological work in the 1790s, which established the reality of extinction, explained this by local catastrophes, after which other fixed species repopulated the affected areas. In Britain, geologists adapted this idea into "diluvial theory" which proposed repeated worldwide annihilation and creation of new fixed species adapted to a changed environment, initially identifying the most recent catastrophe as the biblical flood.

19th Century

From 1830 to 1833 Charles Lyell's multi-volume *Principles of Geology* was published. The work's subtitle was "An attempt to explain the former changes of the Earth's surface by reference to causes now in operation". He drew his explanations from field studies conducted directly before he went to work on the founding geology text, and developed Hutton's idea that the earth was shaped entirely by slow-moving forces still in operation today, acting over a very long period of time. The terms *uniformitarianism* for this idea, and *catastrophism* for the opposing viewpoint, were coined by William Whewell in a review of Lyell's book. *Principles of Geology* was the most influential geological work in the middle of the 19th century.

Lyell's Uniformitarianism

According to Reijer Hooykaas (1963), Lyell's uniformitarianism is a family of four related propositions, not a single idea:

- Uniformity of law – the laws of nature are constant across time and space.

- Uniformity of methodology – the appropriate hypotheses for explaining the geological past are those with analogy today.

- Uniformity of kind – past and present causes are all of the same kind, have the same energy, and produce the same effects.

- Uniformity of degree – geological circumstances have remained the same over time.

None of these connotations requires another, and they are not all equally inferred by uniformitarians.

Gould explained Lyell's propositions in *Time's Arrow, Time's Cycle* (1987), stating that Lyell conflated two different types of propositions: a pair of methodological assumptions with a pair of substantive hypotheses. The four together make up Lyell's uniformitarianism.

Methodological Assumptions

The two methodological assumptions below are accepted to be true by the majority of scientists and geologists. Gould claims that these philosophical propositions must be assumed before you can proceed as a scientist doing science. "You cannot go to a rocky outcrop and observe either the constancy of nature's laws or the working of unknown processes. It works the other way around." You first assume these propositions and "then you go to the outcrop."

- Uniformity of law across time and space: Natural laws are constant across space and time.

The axiom of uniformity of law is necessary in order for scientists to extrapolate (by inductive inference) into the unobservable past. The constancy of natural laws must be assumed in the study of the past; else we cannot meaningfully study it.

- Uniformity of process across time and space: Natural processes are constant across time and space.

Though similar to uniformity of law, this second *a priori* assumption, shared by the vast majority of scientists, deals with geological causes, not physico-chemical laws. The past is to be explained by processes acting currently in time and space rather than inventing extra esoteric or unknown processes *without good reason*, otherwise known as parsimony or Occam's razor.

Substantive Hypotheses

The substantive hypotheses were controversial and, in some cases, accepted by few. These hypotheses are judged true or false on empirical grounds through scientific observation and repeated experimental data. This is in contrast with the previous two philosophical assumptions that come before one can do science and so cannot be tested or falsified by science.

- Uniformity of rate across time and space: Change is typically slow, steady, and gradual.

Uniformity of rate (or gradualism) is what most people (including geologists) think of when they hear the word "uniformitarianism," confusing this hypothesis with the entire definition. As late as 1990, Lemon, in his textbook of stratigraphy, affirmed that "The uniformitarian view of earth history held that all geologic processes proceed continuously and at a very slow pace."

Gould explained Hutton's view of uniformity of rate; mountain ranges or grand canyons are built by accumulation of nearly insensible changes added up through vast time. Some major events such as floods, earthquakes, and eruptions, do occur. But these catastrophes are strictly local. They neither occurred in the past, nor shall happen in the future, at any greater frequency or extent than they display at present. In particular, the whole earth is never convulsed at once.

- Uniformity of state across time and space: Change is evenly distributed throughout space and time.

The uniformity of state hypothesis implies that throughout the history of our earth there is no progress in any inexorable direction. The planet has almost always looked and behaved as it does now. Change is continuous, but leads nowhere. The earth is in balance: a dynamic steady state.

20th Century

Stephen Jay Gould's first scientific paper, *Is uniformitarianism necessary?* (1965), reduced these four interpretations to two. He dismissed the first principle, which asserted spatial and temporal invariance of natural laws, as no longer an issue of debate. He rejected the third (uniformity of rate) as an unjustified limitation on scientific inquiry, as it constrains past geologic rates and conditions to those of the present. So, Lyellian uniformitarianism was unnecessary.

Modern Uniformitarianism Includes Periodic Catastrophes

Uniformitarianism was originally proposed in contrast to catastrophism, which states that the distant past "consisted of epochs of paroxysmal and catastrophic action interposed between periods of comparative tranquility" Especially in the late 19th and early 20th centuries, most geologists took this interpretation to mean that catastrophic events are not important in geologic time; one example of this is the debate of the formation of the Channeled Scablands due to the catastrophic Missoula glacial outburst floods. An important result of this debate and others was the re-clarification that, while the same principles operate in geologic time, catastrophic events that are infrequent on human time-scales can have important consequences in geologic history. Derek Ager has noted that "geologists do not deny uniformitarianism in its true sense, that is to say, of interpreting the past by means of the processes that are seen going on at the present day, so long as we remember that the periodic catastrophe is one of those processes. Those periodic catastrophes make more showing in the stratigraphical record than we have hitherto assumed."

Even Charles Lyell thought that ordinary geological processes would cause Niagara Falls to move upstream to Lake Erie within 10,000 years, leading to catastrophic flooding of a large part of North America.

Modern geologists do not apply uniformitarianism in the same way as Lyell. They question if rates of processes were uniform through time and only those values measured during the history of geology are to be accepted. The present may not be a long enough key to penetrate the deep lock of the past. Geologic processes may have been active at different rates in the past that humans have not observed. "By force of popularity, uniformity of rate has persisted to our present day. For more than a century, Lyell's rhetoric conflating axiom with hypotheses has descended in unmodified form. Many geologists have been stifled by the belief that proper methodology includes an a priori commitment to gradual change, and by a preference for explaining large-scale phenomena as the concatenation of innumerable tiny changes."

The current consensus is that Earth's history is a slow, gradual process punctuated by occasional natural catastrophic events that have affected Earth and its inhabitants. In practice it is reduced from Lyell's conflation to simply the two philosophical assumptions. This is also known as the principle of geological actualism, which states that all past geological action was like all present geological action. The principle of actualism is the cornerstone of paleoecology.

Principle of Original Horizontality

The Principle of Original Horizontality states that layers of sediment are originally deposited horizontally under the action of gravity . It is a relative dating technique. The principle is important to the analysis of folded and tilted strata. It was first proposed by the Danish geological pioneer Nicholas Steno (1638–1686).

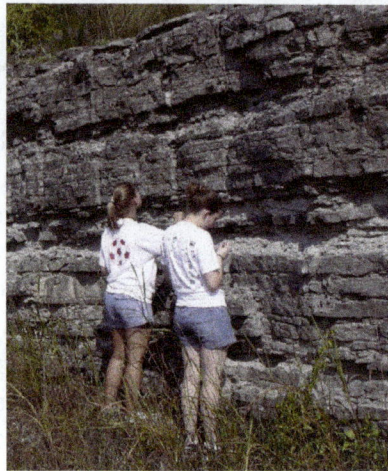

A stratigraphic section of Ordovician rock exposed in central Tennessee, USA. The sediments composing these rocks were formed in an ocean and deposited in horizontal layers.

The Permian through Jurassic stratigraphy of the Colorado Plateau area of southeastern Utah is a great example of Original Horizontality. These strata make up much of the famous prominent rock formations in widely spaced protected areas such as Capitol Reef National Park and Canyonlands National Park. From top to bottom: Rounded tan domes of the Navajo Sandstone, layered red Kayenta Formation, cliff-forming, vertically jointed, red Wingate Sandstone, slope-forming, purplish Chinle Formation, layered, lighter-red Moenkopi Formation, and white, layered Cutler Formation sandstone. Picture from Glen Canyon National Recreation Area, Utah.

From these observations is derived the conclusion that the Earth has not been static and that great forces have been at work over long periods of time, further leading to the conclusions of the science of plate tectonics; that movement and collisions of large plates of the Earth's crust is the cause of folded strata.

As one of Steno's Laws, the Principle of Original Horizontality served well in the nascent days of geological science. However, it is now known that not all sedimentary layers are deposited purely horizontally. For instance, coarser grained sediments such as sand may be deposited at angles of up to 15 degrees, held up by the internal friction between grains which prevents them slumping to a lower angle without additional reworking or effort. This is known as the angle of repose, and a prime example is the surface of sand dunes.

Similarly, sediments may drape over a pre-existing inclined surface: these sediments are usually deposited conformably to the pre-existing surface. Also sedimentary beds may pinch out along strike, implying that slight angles existed during their deposition. Thus the Principle of Original Horizontality is widely, but not universally, applicable in the study of sedimentology, stratigraphy and structural geology.

Law of Superposition

The law of superposition is an axiom that forms one of the bases of the sciences of geology, archaeology, and other fields dealing with geological stratigraphy. In its plainest form, it states that in undeformed stratigraphic sequences, the oldest strata will be at the bottom of the sequence. This is important to stratigraphic dating, which assumes that the law of superposition holds true and that an object cannot be older than the materials of which it is composed. The law was first proposed in the 17th century by the Danish scientist Nicolas Steno.

Layer upon layer of rocks on north shore of Isfjord, Svalbard, Norway. Since there is no overturning, the rock at the bottom is older than the rock on the top by the Law of Superposition.

Archaeological Considerations

Superposition in archaeology and especially in stratification use during excavation is slightly different as the processes involved in laying down archaeological strata are somewhat different from geological processes. Man made intrusions and activity in the archaeological record need not form chronologically from top to bottom or be deformed from the horizontal as natural strata are by equivalent processes. Some archaeological strata (often termed as contexts or layers) are created by undercutting previous strata. An example would be that the silt back-fill of an underground drain would form some time after the ground immediately above it. Other examples of non vertical superposition would be modifications to standing structures such as the creation of new doors and windows in a wall. Superposition in archaeology requires a degree of interpretation to correctly identify chronological sequences and in this sense superposition in archaeology is more dynamic and multi-dimensional.

Principle of Lateral Continuity

The principle of lateral continuity states that layers of sediment initially extend laterally in all directions; in other words, they are laterally continuous. As a result, rocks that are otherwise similar, but are now separated by a valley or other erosional feature, can be assumed to be originally continuous.

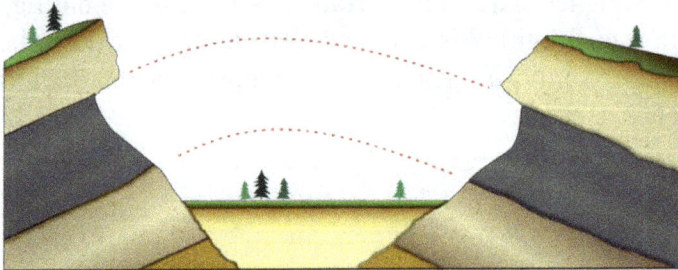

Schematic representation of the principle of lateral continuity.

Layers of sediment do not extend indefinitely; rather, the limits can be recognized and are controlled by the amount and type of sediment available and the size and shape of the sedimentary basin. As long as sediment is transported to an area, it will eventually be deposited. However, as the amount of material lessens away from the source, the layer of that material will become thinner.

Often, coarser-grained material can no longer be transported to an area because the transporting medium has insufficient energy to carry it to that location. In its place, the particles that settle from the transporting medium will be finer-grained, and there will be a lateral transition from coarser- to finer-grained material. The lateral variation in sediment within a stratum is known as sedimentary facies.

If sufficient sedimentary material is available, it will be deposited up to the limits of the sedimentary basin. Often, the sedimentary basin is within rocks that are very different from the sediments that are being deposited. In those cases, the lateral limits of the sedimentary layer will be marked by an abrupt change in rock type.

Principle of Faunal Succession

The principle of faunal succession, also known as the law of faunal succession, is based on the observation that sedimentary rock strata contain fossilized flora and fauna, and that these fossils succeed each other vertically in a specific, reliable order that can be identified over wide horizontal distances. A fossilized Neanderthal bone will never be found in the same stratum as a fossilized Megalosaurus, for example, because neanderthals and megalosaurs lived during different geological periods, separated by many millions of years. This allows for strata to be identified and dated by the fossils found within.

This principle, which received its name from the English geologist William Smith, is of great importance in determining the relative age of rocks and strata. The fossil content of rocks together with the law of superposition helps to determine the time sequence in which sedimentary rocks were laid down.

Evolution explains the observed faunal and floral succession preserved in rocks. Faunal succession was documented by Smith in England during the first decade of the 19th century, and concurrently in France by Cuvier (with the assistance of the mineralogist Alexandre Brongniart). Archaic biological features and organisms are succeeded in the fossil record by more modern versions. For instance, paleontologists investigating the evolution of birds predicted that feathers would first be seen in primitive forms on flightless predecessor organisms such as feathered dinosaurs. This is precisely what has been discovered in the fossil record: simple feathers, incapable of supporting flight, are succeeded by increasingly large and complex feathers.

In practice, the most useful diagnostic species are those with the fastest rate of species turnover and the widest distribution; their study is termed biostratigraphy, the science of dating rocks by using the fossils contained within them. In Cenozoic strata, fossilized tests of foraminifera are often used to determine faunal succession on a refined scale, each biostratigraphic unit (biozone) being a geological stratum that is defined on the basis of its characteristic fossil taxa. An outline microfaunal zonal scheme based on both foraminifera and ostracoda was compiled by M. B. Hart (1972).

Simply, the earlier fossil life forms are simpler than more recent forms, and more recent forms are most similar to existing forms (principle of faunal succession).

References

- Bowler, Peter J. (2003). Evolution: The History of an Idea (3rd ed.). University of California Press. ISBN 0-520-23693-9.

- Levin, H.L. (2009). The Earth Through Time (9 ed.). John Wiley and Sons. p. 15. ISBN 978-0-470-38774-0. Retrieved 28 November 2010.

- Principles of Archaeological Stratigraphy. 40 figs. 1 pl. 136 pp. London & New York: Academic Press ISBN 0-12-326650-5

- Winchester, Simon (2001), The Map that Changed the World: William Smith and the Birth of Modern Geology, New York: HarperCollins, pp. 59–91, ISBN 0-06-093180-9

4

Sub-fields of Geology

Sub-fields of geology include economic geology, engineering geology, mining, geotechnical engineering, plate tectonics, hydrogeology, planetary geology, biochemistry etc. Economic geology is the study of Earth's materials, mostly the ones that are used for economical or industrial purposes. This text helps the reader in developing an in-depth understanding of all the sub-fields of geology.

Economic Geology

An open pit uranium mine in Namibia

Economic geology is concerned with earth materials that can be used for economic and/or industrial purposes. These materials include precious and base metals, non-metallic minerals, construction-grade stone, petroleum minerals, coal, and water. The term commonly refers to metallic mineral deposits and mineral resources. The techniques employed by other earth science disciplines (such as geochemistry, mineralogy, geophysics, petrology and structural geology) might all be used to understand, describe, and exploit an ore deposit.

Economic geology is studied and practiced by geologists. However it is of prime interest to investment bankers, stock analysts and other professions such as engineers, environmental scientists, and conservationists because of the far-reaching impact that extractive industries have on society, the economy, and the environment.

Purpose of Studies

The purpose of studies of the subject is as follows:

1. The subject of *Economic geology* is aimed to provide a detailed description of economic and geologic materials, the number of which is estimated to be around 200. Besides the detailed description of these materials, the subject also discusses their proper use and development. The foremost duty of economic geologists is to determine the suitability of a mineral given a particular industry.

2. From an economic perspective, a mineral's reserve is limited, its occurrence variable, and its supply non-replenishable; once a mineral is fully extracted, its reserve is exhausted. These concerns are what motivate an economic study of geology. A wise economic geologist prepares a plan for the mineral deposit according to proper utilization before extraction.

Mineral Resources

Mineral resources are concentrations of minerals significant for current and future societal needs. Ore is classified as mineralization economically and technically feasible for extraction. Not all mineralization meets these criteria for various reasons. The specific categories of mineralization in an economic sense are:

*   *Mineral occurrences* or prospects of geological interest but not necessarily economic interest

*   *Mineral resources* include those potentially economically and technically feasible and those that are not

*   *Ore reserves*, which must be economically and technically feasible to extract

Ore Geology

Geologists are involved in the study of ore deposits, which includes the study of ore genesis and the processes within the Earth's crust that form and concentrate ore minerals into economically viable quantities.

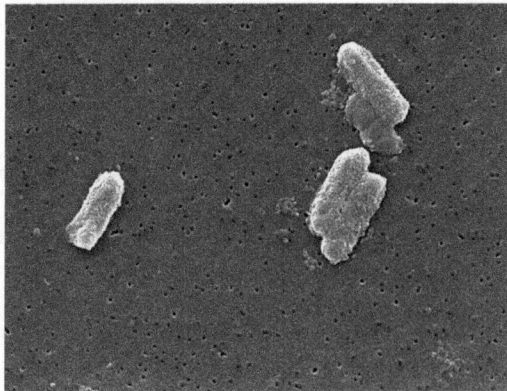

Citrobacter species can have concentrations of uranium in their bodies 300 times higher than in the surrounding environment.

Study of metallic ore deposits involves the use of structural geology, geochemistry, the study of metamorphism and its processes, as well as understanding metasomatism and other processes related to ore genesis.

Ore deposits are delineated by mineral exploration, which uses geochemical prospecting, drilling and resource estimation via geostatistics to quantify economic ore bodies. The ultimate aim of this process is mining.

Coal and Petroleum Geology

The study of sedimentology is of prime importance to the delineation of economic reserves of petroleum and coal energy resources.

Mud log in process, a common way to study the lithology when drilling oil wells.

Petroleum Geology

Petroleum geology is the study of origin, occurrence, movement, accumulation, and exploration of hydrocarbon fuels. It refers to the specific set of geological disciplines that are applied to the search for hydrocarbons (oil exploration).

Sedimentary Basin Analysis

Petroleum geology is principally concerned with the evaluation of seven key elements in sedimentary basins:

- Source
- Reservoir
- Seal
- Trap
- Timing

- Maturation

- Migration

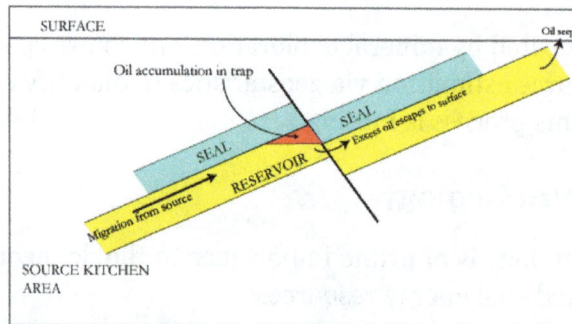

A structural trap, where a fault has juxtaposed a porous and permeable reservoir against an impermeable seal. Oil (shown in red) accumulates against the seal, to the depth of the base of the seal. Any further oil migrating in from the source will escape to the surface and seep.

In general, all these elements must be assessed via a limited 'window' into the subsurface world, provided by one (or possibly more) exploration wells. These wells present only a 1-dimensional segment through the Earth and the skill of inferring 3-dimensional characteristics from them is one of the most fundamental in petroleum geology. Recently, the availability of inexpensive, high quality 3D seismic data (from reflection seismology) and data from various electromagnetic geophysical techniques (such as Magnetotellurics) has greatly aided the accuracy of such interpretation. The following section discusses these elements in brief.

Evaluation of the source uses the methods of geochemistry to quantify the nature of organic-rich rocks which contain the precursors to hydrocarbons, such that the type and quality of expelled hydrocarbon can be assessed.

The reservoir is a porous and permeable lithological unit or set of units that holds the hydrocarbon reserves. Analysis of reservoirs at the simplest level requires an assessment of their porosity (to calculate the volume of *in situ* hydrocarbons) and their permeability (to calculate how easily hydrocarbons will flow out of them). Some of the key disciplines used in reservoir analysis are the fields of structural analysis, stratigraphy, sedimentology, and reservoir engineering.

The seal, or *cap* rock, is a unit with low permeability that impedes the escape of hydrocarbons from the reservoir rock. Common seals include evaporites, chalks and shales. Analysis of seals involves assessment of their thickness and extent, such that their effectiveness can be quantified.

The trap is the stratigraphic or structural feature that ensures the juxtaposition of reservoir and seal such that hydrocarbons remain trapped in the subsurface, rather than escaping (due to their natural buoyancy) and being lost.

Analysis of maturation involves assessing the thermal history of the source rock in order to make predictions of the amount and timing of hydrocarbon generation and expulsion.

Finally, careful studies of migration reveal information on how hydrocarbons move from source to reservoir and help quantify the source (or *kitchen*) of hydrocarbons in a particular area.

Major Subdisciplines in Petroleum Geology

Several major subdisciplines exist in petroleum geology specifically to study the seven key elements discussed above.

Source Rock Analysis

In terms of source rock analysis,several facts need to be established. Firstly, the question of whether there actually *is* any source rock in the area must be answered. Delineation and identification of potential source rocks depends on studies of the local stratigraphy, palaeogeography and sedimentology to determine the likelihood of organic-rich sediments having been deposited in the past.

If the likelihood of there being a source rock is thought to be high, the next matter to address is the state of thermal maturity of the source, and the timing of maturation. Maturation of source rocks depends strongly on tem-perature, such that the majority of oil generation occurs in the 60° to 120°C range. Gas generation starts at similar temperatures, but may continue up beyond this range, perhaps as high as 200°C. In order to determine the likelihood of oil/gas generation, therefore, the thermal history of the source rock must be calculated. This is performed with a combination of geochemical analysis of the source rock (to determine the type of kerogens present and their maturation characteristics) and basin modelling methods, such as back-stripping, to model the thermal gradient in the sedimentary column.

Basin Analysis

A full scale basin analysis is usually carried out prior to defining leads and prospects for future drilling. This study tackles the petroleum system and studies source rock (presence and quality); burial history; maturation (timing and volumes); migration and focus; and potential regional seals and major reservoir units (that define carrier beds). All these elements are used to investigate where potential hydrocarbons might migrate towards. Traps and potential leads and prospects are then defined in the area that is likely to have received hydrocarbons.

Exploration Stage

Although a basin analysis is usually part of the first study a company conducts prior to moving into an area for future exploration, it is also sometimes conducted during the exploration phase. Exploration geology comprises all the activities and studies necessary for finding new hydrocarbon occurrence. Usually seismic (or 3D seismic) studies are shot, and old exploration data (seismic lines, well logs, reports) are used to expand upon the new studies. Sometimes gravity and magnetic studies are conducted, and oil seeps and spills are mapped to find potential areas for hydrocarbon occurrences. As soon as a significant hydrocarbon occurrence is found by an exploration- or wild-cat-well the appraisal stage starts.

Appraisal Stage

The Appraisal stage is used to delineate the extent of the discovery. Hydrocarbon reservoir properties, connectivity, hydrocarbon type and gas-oil and oil-water contacts are determined to calculate potential recoverable volumes. This is usually done by drilling more appraisal wells around the initial exploration well. Production tests may also give insight in reservoir pressures and connectivity. Geochemical and petrophysical analysis gives information on the type (viscosity, chemistry, API, carbon content, etc.) of the hydrocarbon and the nature of the reservoir (porosity, permeability, etc.).

Production Stage

After a hydrocarbon occurrence has been discovered and appraisal has indicated it is a commercial find the production stage is initiated. This stage focuses on extracting the hydrocarbons in a controlled way (without damaging the formation, within commercial favorable volumes, etc.). Production wells are drilled and completed in strategic positions. 3D seismic is usually available by this stage to target wells precisely for optimal recovery. Sometimes enhanced recovery (steam injection, pumps, etc.) is used to extract more hydrocarbons or to redevelop abandoned fields.

Reservoir Analysis

The existence of a reservoir rock (typically, sandstones and fractured limestones) is determined through a combination of regional studies (i.e. analysis of other wells in the area), stratigraphy and sedimentology (to quantify the pattern and extent of sedimentation) and seismic interpretation. Once a possible hydrocarbon reservoir is identified, the key physical characteristics of a reservoir that are of interest to a hydrocarbon explorationist are its bulk rock volume, net-to-gross ratio, porosity and permeability.

Bulk rock volume, or the gross rock volume of rock above any hydrocarbon-water contact, is determined by mapping and correlating sedimentary packages. The net-to-gross ratio, typically estimated from analogues and wireline logs, is used to calculate the proportion of the sedimentary packages that contains reservoir rocks. The bulk rock volume multiplied by the net-to-gross ratio gives the net rock volume of the reservoir. The net rock volume multiplied by porosity gives the total hydrocarbon pore volume i.e. the volume within the sedimentary package that fluids (importantly, hydrocarbons and water) can occupy. The summation of these volumes for a given exploration prospect will allow explorers and commercial analysts to determine whether a prospect is financially viable.

Traditionally, porosity and permeability were determined through the study of drilling samples, analysis of cores obtained from the wellbore, examination of contiguous parts of the reservoir that outcrop at the surface and by the technique of formation evaluation using wireline tools passed down the well itself. Modern advances in seismic data acquisition and processing have meant that seismic attributes of subsurface rocks are readily available and can be used to infer physical/sedimentary properties of the rocks themselves.

Engineering Geology

Engineering geology is the application of the geological sciences to engineering study for the purpose of assuring that the geological factors regarding the location, design, construction, operation and maintenance of engineering works are recognized and accounted for. Engineering geologists provide geological and geotechnical recommendations, analysis, and design associated with human development and various types of structures. The realm of the engineering geologist is essentially in the area of earth-structure interactions, or investigation of how the earth or earth processes impact human made structures and human activities.

An engineering geologist logging rock core in the field, Western Australia.

Engineering geology studies may be performed during the planning, environmental impact analysis, civil or structural engineering design, value engineering and construction phases of public and private works projects, and during post-construction and forensic phases of projects. Works completed by engineering geologists include; geological hazard assessments, geotechnical, material properties, landslide and slope stability, erosion, flooding, dewatering, and seismic investigations, etc. Engineering geology studies are performed by a geologist or engineering geologist that is educated, trained and has obtained experience related to the recognition and interpretation of natural processes, the understanding of how these processes impact human made structures (and vice versa), and knowledge of methods by which to mitigate against hazards resulting from adverse natural or human made conditions. The principal objective of the engineering geologist is the protection of life and property against damage caused by various geological conditions.

The practice of engineering geology is also very closely related to the practice of geological engineering and geotechnical engineering. If there is a difference in the content of the disciplines, it mainly lies in the training or experience of the practitioner.

History

Although the study of geology has been around for centuries, at least in its modern form, the science and practice of engineering geology only commenced as a recognized discipline until the late 19th and early 20th centuries. The first book titled Engineering Geology was published in 1880 by William Penning. In the early 20th century Charles Berkey, an American trained geologist who was considered the first American engineering geologist, worked on several water-supply projects for New York City, then later worked on the Hoover dam and a multitude of other engineering projects. The first American engineering geology textbook was written in 1914 by Ries and Watson. In 1921 Reginald W. Brock, the first Dean of Applied Science at the University of British Columbia, started the first undergraduate and graduate degree programs in Geological Engineering, noting that students with an engineering foundation made first-class practicing geologists. In 1925, Karl Terzaghi, an Austrian trained engineer and geologist, published the first text in Soil Mechanics (in German). Terzaghi is known as the parent of soil mechanics, but also had great interest in geology; Terzaghi considered soil mechanics to be a sub-discipline of engineering geology. In 1929, Terzaghi, along with Redlich and Kampe, published their own Engineering Geology text (also in German).

The need for geologist on engineering works gained worldwide attention in 1928 with the failure of the St. Francis Dam in California and the death of 426 people. More engineering failures which occurred the following years also prompted the requirement for engineering geologists to work on large engineering projects.

In 1951, one of the earliest definitions of the "Engineering Geologist" or "Professional Engineering Geologist" was provided by the Executive Committee of the Division on Engineering Geology of the Geological Society of America.

The Practice

One of the most important roles as an engineering geologist is the interpretation of landforms and earth processes to identify potential geologic and related man-made hazards that may have a great impact on civil structures and human development.The background in geology provides the engineering geologist with an understanding of how the earth works, which is crucial minimizing earth related hazards. Most engineering geologists also have graduate degrees where they have gained specialized education and training in soil mechanics, rock mechanics, geotechnics, groundwater, hydrology, and civil design. These two aspects of the engineering geologists' education provides them with a unique ability to understand and mitigate for hazards associated with earth-structure interactions.

Scope of Studies

Engineering geology investigation and studies may be performed:

- for residential, commercial and industrial developments;

- for governmental and military installations;

- for public works such as a stormwater drainage system, power plant, wind turbine, transmission line, sewage treatment plant, water treatment plant, pipeline (aqueduct, sewer, outfall), tunnel, trenchless construction, canal, dam, reservoir, building foundation, railroad, transit, highway, bridge, seismic retrofit, power generation facility, airport and park;

- for mine and quarry developments, mine tailing dam, mine reclamation and mine tunneling;

- for wetland and habitat restoration programs;

- for government, commercial, or industrial hazardous waste remediation sites;

- for coastal engineering, sand replenishment, bluff or sea cliff stability, harbor, pier and waterfront development;

- for offshore outfall, drilling platform and sub-sea pipeline, sub-sea cable; and

- for other types of facilities.

Geohazards and Adverse Geological Conditions

Typical geologic hazards or other adverse conditions evaluated and mitigated by an engineering geologist include:

- fault rupture on seismically active faults ;

- seismic and earthquake hazards (ground shaking, liquefaction, lurching, lateral spreading, tsunami and seiche events);

- landslide, mudflow, rockfall, debris flow, and avalanche hazards ;

- unstable slopes and slope stability;

- erosion;

- slaking and heave of geologic formations, such as frost heaving;

- ground subsidence (such as due to ground water withdrawal, sinkhole collapse, cave collapse, decomposition of organic soils, and tectonic movement);

- volcanic hazards (volcanic eruptions, hot springs, pyroclastic flows, debris flow, debris avalanche, gas emissions, volcanic earthquakes);

- non-rippable or marginally rippable rock requiring heavy ripping or blasting;

- weak and collapsible soils, foundation bearing failures;

- shallow ground water/seepage; and

- other types of geologic constraints.

An engineering geologist or geophysicist may be called upon to evaluate the excavatability (i.e. rippability) of earth (rock) materials to assess the need for pre-blasting during earthwork construction, as well as associated impacts due to vibration during blasting on projects.

Soil and Rock Mechanics

Soil mechanics is a discipline that applies principles of engineering mechanics, e.g. kinematics, dynamics, fluid mechanics, and mechanics of material, to predict the mechanical behavior of soils. Rock mechanics is the theoretical and applied science of the mechanical behaviour of rock and rock masses; it is that branch of mechanics concerned with the response of rock and rock masses to the force-fields of their physical environment. The fundamental processes are all related to the behaviour of porous media. Together, soil and rock mechanics are the basis for solving many engineering geology problems.

Methods and Reporting

The methods used by engineering geologists in their studies include

- geologic field mapping of geologic structures, geologic formations, soil units and hazards;

- the review of geologic literature, geologic maps, geotechnical reports, engineering plans, environmental reports, stereoscopic aerial photographs, remote sensing data, Global Positioning System (GPS) data, topographic maps and satellite imagery;

- the excavation, sampling and logging of earth/rock materials in drilled borings, backhoe test pits and trenches, fault trenching, and bulldozer pits;

- geophysical surveys (such as seismic refraction traverses, resistivity surveys, ground penetrating radar (GPR) surveys, magnetometer surveys, electromagnetic surveys, high-resolution sub-bottom profiling, and other geophysical methods);

- deformation monitoring as the systematic measurement and tracking of the alteration in the shape or dimensions of an object as a result of the application of stress to it manually or with an automatic deformation monitoring system; and

- other methods.

The fieldwork is typically culminated in analysis of the data and the preparation of an engineering geologic report, geotechnical report or design brief, fault hazard or seismic hazard report, geophysical report, ground water resource report or hydrogeologic report. The engineering geology report can also be prepared in conjunction with a geotechnical report, but commonly provides the same geotechnical analysis and design recommendations that would be presented in a geotechnical report. An engineering geology report describes the objectives, methodology, references cited, tests performed, findings and recommendations for development and detailed design of engineering works. Engineering geologists also provide geologic data on topographic maps, aerial photographs, geologic maps, Geographic Information System (GIS) maps, or other map bases.

Mining

Mining is the extraction of valuable minerals or other geological materials from the earth from an orebody, lode, vein, seam, reef or placer deposits which forms the mineralized package of economic interest to the miner.

Surface coal mining

Simplified world active active mining map

Sulfur miner with 90 kg of sulfur carried from the floor of the Ijen Volcano (2015)

Ores recovered by mining include metals, coal, oil shale, gemstones, limestone, dimension stone, rock salt, potash, gravel, and clay. Mining is required to obtain any material that cannot be grown through agricultural processes, or created artificially in a laboratory or factory. Mining in a wider sense includes extraction of any non-renewable resource such as petroleum, natural gas, or even water.

Mining of stones and metal has been a human activity since pre-historic times. Modern mining processes involve prospecting for ore bodies, analysis of the profit potential of a proposed mine, extraction of the desired materials, and final reclamation of the land after the mine is closed.

Mining operations usually create a negative environmental impact, both during the mining activity and after the mine has closed. Hence, most of the world's nations have passed regulations to decrease the impact. Worker safety has long been a concern as well, and modern practices have significantly improved safety in mines.

Levels of metals recycling are generally low. Unless future end-of-life recycling rates are stepped up, some rare metals may become unavailable for use in a variety of consumer products. Due to the low recycling rates, some landfills now contain higher concentrations of metal than mines themselves.

History

Prehistoric Mining

Since the beginning of civilization, people have used stone, ceramics and, later, metals found close to the Earth's surface. These were used to make early tools and weapons; for example, high quality flint found in northern France and southern England was used to create flint tools. Flint mines have been found in chalk areas where seams of the stone were followed underground by shafts and galleries. The mines at Grimes Graves are especially famous, and like most other flint mines, are Neolithic in origin (ca 4000 BC-ca 3000 BC). Other hard rocks mined or collected for axes included the greenstone of the Langdale axe industry based in the English Lake District.

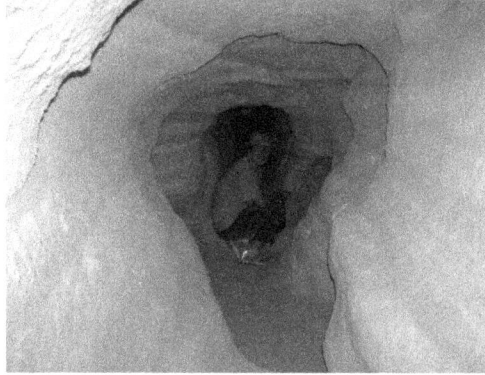

Chalcolithic copper mine in Timna Valley, Negev Desert

The oldest known mine on archaeological record is the "Lion Cave" in Swaziland, which radiocarbon dating shows to be about 43,000 years old. At this site Paleolithic humans mined hematite to make the red pigment ochre. Mines of a similar age in Hungary are believed to be sites where Neanderthals may have mined flint for weapons and tools.

Ancient Egypt

Ancient Egyptians mined malachite at Maadi. At first, Egyptians used the bright green malachite stones for ornamentations and pottery. Later, between 2613 and 2494 BC, large building projects required expeditions abroad to the area of Wadi Maghareh in order to secure minerals and other resources not available in Egypt itself. Quarries for turquoise and copper were also found at Wadi Hamamat, Tura, Aswan and various other Nubian sites on the Sinai Peninsula and at Timna.

Mining in Egypt occurred in the earliest dynasties. The gold mines of Nubia were among the largest and most extensive of any in Ancient Egypt. These mines are described by the Greek author Diodorus Siculus, who mentions fire-setting as one method used to break down the hard rock holding the gold. One of the complexes is shown in one of the earliest known maps. The miners crushed the ore and ground it to a fine powder before washing the powder for the gold dust.

Ancient Greek and Roman Mining

Mining in Europe has a very long history. Examples include the silver mines of Laurium, which helped support the Greek city state of Athens. Despite the mine having over 20,000 slaves working in them, the technology was essentially identical to their Bronze Age predecessors. Other mines, such as on the island of Thassos, had marble quarried by the Parians after having arrived in the 7th Century BC. The marble was shipped away and was later found by archaeologists to have been used in buildings including the tomb of Amphipolis. Philip II of Macedon, the father of Alexander the Great, captured the gold mines of Mount Pangeo in 357 BC to fund his military campaigns. He also captured gold mines in Thrace for minting coinage, eventually producing 26 tons per year.

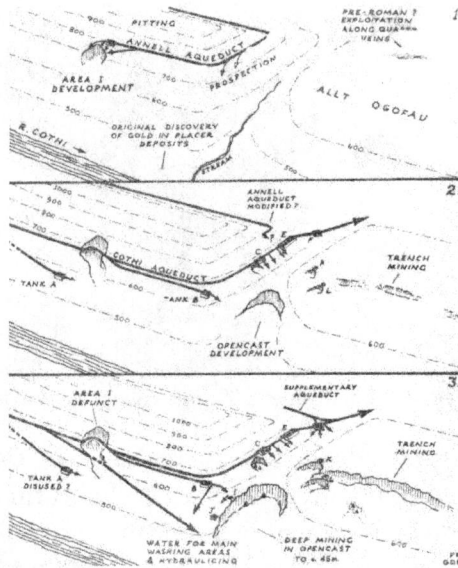

Ancient Roman development of the Dolaucothi Gold Mines, Wales

However, it was the Romans who developed large scale mining methods, especially the use of large volumes of water brought to the minehead by numerous aqueducts. The water was used for a variety of purposes, including removing overburden and rock debris, called hydraulic mining, as well as washing comminuted, or crushed, ores and driving simple machinery.

The Romans used hydraulic mining methods on a large scale to prospect for the veins of ore, especially a now obsolete form of mining known as hushing. This method involved building numerous aqueducts to supply water to the minehead where it was stored in large reservoirs and tanks. When a full tank was opened, the flood of water sluiced away the overburden to expose the bedrock underneath and any gold veins. The rock was then worked upon by fire-setting to heat the rock, which would be quenched with a stream of water. The resulting thermal shock cracked the rock, enabling it to be removed, aided by further streams of water from the overhead tanks. The Roman miners used similar methods to work cassiterite deposits in Cornwall and lead ore in the Pennines.

The methods had been developed by the Romans in Spain in 25 AD to exploit large alluvial gold deposits, the largest site being at Las Medulas, where seven long aqueducts were built to tap local rivers and to sluice the deposits. Spain was one of the most important mining regions, but all regions of the Roman Empire were exploited. In Great Britain the natives had mined minerals for millennia, but after the Roman conquest, the scale of the operations increased dramatically, as the Romans needed Britannia's resources, especially gold, silver, tin, and lead.

Roman techniques were not limited to surface mining. They followed the ore veins underground once opencast mining was no longer feasible. At Dolaucothi they stoped out the veins, and drove adits through barren rock to drain the stopes. The same adits were also used to ventilate the workings, especially important when fire-setting was used. At other parts of the site, they penetrated the water table and dewatered the mines using several kinds of machines, especially reverse overshot water-wheels. These were used extensively in the copper mines at Rio Tinto in Spain, where one sequence comprised 16 such wheels arranged in pairs, and lifting water about 80 feet (24 m). They were worked as treadmills with miners standing on the top slats. Many examples of such devices have been found in old Roman mines and some examples are now preserved in the British Museum and the National Museum of Wales.

Medieval Europe

Mining as an industry underwent dramatic changes in medieval Europe. The mining industry in the early Middle Ages was mainly focused on the extraction of copper and iron. Other precious metals were also used, mainly for gilding or coinage. Initially, many metals were obtained through open-pit mining, and ore was primarily extracted from shallow depths, rather than through deep mine shafts. Around the 14th century, the growing use of weapons, armour, stirrups, and horseshoes greatly increased the demand for iron. Medieval knights, for example, were often laden with up to 100 pounds of plate or chain link armour in addition to swords, lances and other weapons. The overwhelming dependency on iron for military purposes spurred iron production and extraction processes.

Agricola, author of *De Re Metallica*

The silver crisis of 1465 occurred when all mines had reached depths at which the shafts could no longer be pumped dry with the available technology. Although an increased use of bank notes, credit and copper coins during this period did decrease the value of, and dependence on, precious metals, gold and silver still remained vital to the story of medieval mining.

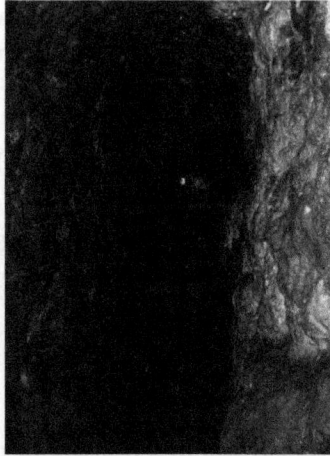

Gallery, 12th to 13th century, Germany

Due to differences in the social structure of society, the increasing extraction of mineral deposits spread from central Europe to England in the mid-sixteenth century. On the continent, all mineral deposits belonged to the crown, and this regalian right was stoutly maintained; but in England, it was pared down to gold and silver (of which there were virtually no deposits) by a judicial decision of 1568 and a law of 1688. England had iron, zinc, copper, lead, and tin ores. Landlords who owned the base metals and coal under their estates were now rendered with a strong inducement to extract these metals or to lease the deposits and collect royalties from mine operators. English, German, and Dutch capital combined to finance extraction and refining. Hundreds of German technicians and skilled workers were brought over; in 1642 a colony of 4,000 foreigners was mining and smelting copper at Keswick in the northwestern mountains.

Use of water power in the form of water mills was extensive. The water mills were employed in crushing ore, raising ore from shafts, and ventilating galleries by powering giant bellows. Black powder was first used in mining in Selmecbánya, Kingdom of Hungary (now Banská Štiavnica, Slovakia) in 1627. Black powder allowed blasting of rock and earth to loosen and reveal ore veins. Blasting was much faster than fire-setting and allowed the mining of previously impenetrable metals and ores. In 1762, the world's first mining academy was established in the same town.

The widespread adoption of agricultural innovations such as the iron plowshare, as well as the growing use of metal as a building material, was also a driving force in the tremendous growth of the iron industry during this period. Inventions like the arrastra were often used by the Spanish to pulverize ore after being mined. This device was powered by animals and used the same principles used for grain threshing.

Much of the knowledge of medieval mining techniques comes from books such as Biringuccio's *De la pirotechnia* and probably most importantly from Georg Agricola's *De re metallica* (1556). These books detail many different mining methods used in German and Saxon mines. One of the prime issues confronting medieval miners (and one which

Agricola explains in detail) was the removal of water from mining shafts. As miners dug deeper to access new veins, flooding became a very real obstacle. The mining industry became dramatically more efficient and prosperous with the invention of mechanical and animal driven pumps.

Classical Philippine Civilization

Mining in the Philippines began around 1000 BC. The early Filipinos worked various mines of gold, silver, copper and iron. Jewels, gold ingots, chains, calombigas and earrings were handed down from antiquity and inherited from their ancestors. Gold dagger handles, gold dishes, tooth plating, and huge gold ornamets were also used. In Laszlo Legeza's "Tantric elements in pre-Hispanic Philippines Gold Art", he mentioned that gold jewelry of Philippine origin was found in Ancient Egypt. According to Antonio Pigafetta, the people of Mindoro possessed great skill in mixing gold with other metals and gave it a natural and perfect appearance that could deceive even the best of silversmiths. The natives were also known for the jewelries made of other precious stones such as carnelian, agate and pearl. Some outstanding examples of Philippine jewelry included necklaces, belts, armlets and rings placed around the waist.

The image of a Maharlika class of the Philippine Society , depicted in Boxer Codex that the Gold used as a form of Jewelry (ca.1400).

The Americas

There are ancient, prehistoric copper mines along Lake Superior, and metallic copper was still found there, near the surface, in colonial times. Indegenous peoples availed themselves of this copper starting at least 5,000 years ago," and copper tools, arrowheads, and other artifacts that were part of an extensive native trade network have been discovered. In addition, obsidian, flint, and other minerals were mined, worked, and traded. Early French explorers who encountered the sites made no use of the metals due to the difficulties of transporting them, but the copper was eventually traded throughout the continent along major river routes.

Lead mining in the upper Mississippi River region of the U.S., 1865.

In the early colonial history of the Americas, "native gold and silver was quickly expro-priated and sent back to Spain in fleets of gold- and silver-laden galleons," the gold and silver originating mostly from mines in Central and South America. Turquoise dated at 700 A.D. was mined in pre-Columbian America; in the Cerillos Mining District in New Mexico, estimates are that "about 15,000 tons of rock had been removed from Mt. Chalchihuitl using stone tools before 1700."

Miners at the Tamarack Mine in Copper Country, Michigan, U.S. in 1905.

Mining in the United States became prevalent in the 19th century, and the General Mining Act of 1872 was passed to encourage mining of federal lands. As with the Cal-ifornia Gold Rush in the mid-19th century, mining for minerals and precious metals, along with ranching, was a driving factor in the Westward Expansion to the Pacific coast. With the exploration of the West, mining camps were established and "expressed a distinctive spirit, an enduring legacy to the new nation;" Gold Rushers would experi-ence the same problems as the Land Rushers of the transient West that preceded them. Aided by railroads, many traveled West for work opportunities in mining. Western cit-ies such as Denver and Sacramento originated as mining towns.

When new areas were explored, it was usually the gold (placer and then load) and then silver that were taken into possession and extracted first. Other metals would often wait for railroads or canals, as coarse gold dust and nuggets do not require smelting and are easy to identify and transport.

Modern Period

In the early 20th century, the gold and silver rush to the western United States also stimulated mining for coal as well as base metals such as copper, lead, and iron. Areas in modern Montana, Utah, Arizona, and later Alaska became predominate suppliers of copper to the world, which was increasingly demanding copper for electrical and households goods. Canada's mining industry grew more slowly than did the United States' due to limitations in transportation, capital, and U.S. competition; Ontario was the major producer of the early 20th century with nickel, copper, and gold.

Meanwhile, Australia experienced the Australian gold rushes and by the 1850s was producing 40% of the world's gold, followed by the establishment of large mines such as the Mount Morgan Mine, which ran for nearly a hundred years, Broken Hill ore deposit (one of the largest zinc-lead ore deposits), and the iron ore mines at Iron Knob. After declines in production, another boom in mining occurred in the 1960s. Now, in the early 21st century, Australia remains a major world mineral producer.

As the 21st century begins, a globalized mining industry of large multinational corporations has arisen. Peak minerals and environmental impacts have also become a concern. Different elements, particularly rare earth minerals, have begun to increase in demand as a result of new technologies.

Mine Development and Lifecycle

The process of mining from discovery of an ore body through extraction of minerals and finally to returning the land to its natural state consists of several distinct steps. The first is discovery of the ore body, which is carried out through prospecting or exploration to find and then define the extent, location and value of the ore body. This leads to a mathematical resource estimation to estimate the size and grade of the deposit.

Schematic of a cut and fill mining operation in hard rock.

This estimation is used to conduct a pre-feasibility study to determine the theoretical economics of the ore deposit. This identifies, early on, whether further investment in estimation and engineering studies is warranted and identifies key risks and areas for further work. The next step is to conduct a feasibility study to evaluate the financial viability, the technical and financial risks, and the robustness of the project.

This is when the mining company makes the decision whether to develop the mine or to walk away from the project. This includes mine planning to evaluate the economically recoverable portion of the deposit, the metallurgy and ore recoverability, marketability and payability of the ore concentrates, engineering concerns, milling and infrastructure costs, finance and equity requirements, and an analysis of the proposed mine from the initial excavation all the way through to reclamation. The proportion of a deposit that is economically recoverable is dependent on the enrichment factor of the ore in the area.

To gain access to the mineral deposit within an area it is often necessary to mine through or remove waste material which is not of immediate interest to the miner. The total movement of ore and waste constitutes the mining process. Often more waste than ore is mined during the life of a mine, depending on the nature and location of the ore body. Waste removal and placement is a major cost to the mining operator, so a detailed characterization of the waste material forms an essential part of the geological exploration program for a mining operation.

Once the analysis determines a given ore body is worth recovering, development begins to create access to the ore body. The mine buildings and processing plants are built, and any necessary equipment is obtained. The operation of the mine to recover the ore begins and continues as long as the company operating the mine finds it economical to do so. Once all the ore that the mine can produce profitably is recovered, reclamation begins to make the land used by the mine suitable for future use.

Mining Techniques

Mining techniques can be divided into two common excavation types: surface mining and sub-surface (underground) mining. Today, surface mining is much more common, and produces, for example, 85% of minerals (excluding petroleum and natural gas) in the United States, including 98% of metallic ores.

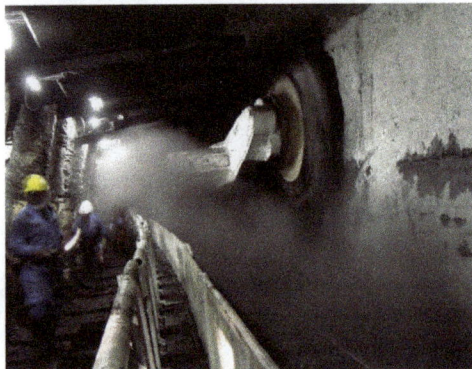
Underground longwall mining.

Targets are divided into two general categories of materials: *placer deposits*, consisting of valuable minerals contained within river gravels, beach sands, and other unconsolidated materials; and *lode deposits*, where valuable minerals are found in veins, in layers, or in mineral grains generally distributed throughout a mass of actual rock. Both types of ore deposit, placer or lode, are mined by both surface and underground methods.

Some mining, including much of the rare earth elements and uranium mining, is done by less-common methods, such as in-situ leaching: this technique involves digging neither at the surface nor underground. The extraction of target minerals by this technique requires that they be soluble, e.g., potash, potassium chloride, sodium chloride, sodium sulfate, which dissolve in water. Some minerals, such as copper minerals and uranium oxide, require acid or carbonate solutions to dissolve.

Surface Mining

Surface mining is done by removing (stripping) surface vegetation, dirt, and, if necessary, layers of bedrock in order to reach buried ore deposits. Techniques of surface mining include: open-pit mining, which is the recovery of materials from an open pit in the ground, quarrying, identical to open-pit mining except that it refers to sand, stone and clay; strip mining, which consists of stripping surface layers off to reveal ore/seams underneath; and mountaintop removal, commonly associated with coal mining, which involves taking the top of a mountain off to reach ore deposits at depth. Most (but not all) placer deposits, because of their shallowly buried nature, are mined by surface methods. Finally, landfill mining involves sites where landfills are excavated and processed.

Garzweiler surface mine, Germany

Underground Mining

Mantrip used for transporting miners within an underground mine

Sub-surface mining consists of digging tunnels or shafts into the earth to reach buried ore deposits. Ore, for processing, and waste rock, for disposal, are brought to the surface through the tunnels and shafts. Sub-surface mining can be classified by the type of access shafts used, the extraction method or the technique used to reach the mineral deposit. Drift mining utilizes horizontal access tunnels, slope mining uses diagonally sloping access shafts, and shaft mining utilizes vertical access shafts. Mining in hard and soft rock formations require different techniques.

Other methods include shrinkage stope mining, which is mining upward, creating a sloping underground room, long wall mining, which is grinding a long ore surface underground, and room and pillar mining, which is removing ore from rooms while leaving pillars in place to support the roof of the room. Room and pillar mining often leads to retreat mining, in which supporting pillars are removed as miners retreat, allowing the room to cave in, thereby loosening more ore. Additional sub-surface mining methods include hard rock mining, which is mining of hard rock (igneous, metamorphic or sedimentary) materials, bore hole mining, drift and fill mining, long hole slope mining, sub level caving, and block caving.

Highwall Mining

Caterpillar Highwall Miner HW300 - Technology Bridging Underground and Open Pit Mining

Highwall mining is another form of surface mining that evolved from auger mining. In Highwall mining, the coal seam is penetrated by a continuous miner propelled by a hydraulic Pushbeam Transfer Mechanism (PTM). A typical cycle includes sumping (launch-pushing forward) and shearing (raising and lowering the cutterhead boom to cut the entire height of the coal seam). As the coal recovery cycle continues, the cutterhead is progressively launched into the coal seam for 19.72 feet (6.01 m). Then, the Pushbeam Transfer Mechanism (PTM) automatically inserts a 19.72-foot (6.01 m) long rectangular Pushbeam (Screw-Conveyor Segment) into the center section of the machine between the Powerhead and the cutterhead. The Pushbeam system can penetrate nearly 1,000 feet (300 m) into the coal seam. One patented Highwall mining systems use augers enclosed inside the Pushbeam that prevent the mined coal from being contaminated by rock debris during the conveyance process. Using a video imaging and/or a gamma ray sensor and/or other Geo-Radar systems like a coal-rock interface detection sensor (CID), the operator can see ahead projection of the seam-rock interface and guide the continuous miner's progress. Highwall mining can produce thousands of tons of coal in contour-strip operations with narrow benches, previously mined areas, trench mine applications and steep-dip seams with controlled water-inflow pump system and/or a gas (inert) venting system.

Machines

Heavy machinery is used in mining to explore and develop sites, to remove and stockpile overburden, to break and remove rocks of various hardness and toughness, to process the ore, and to carry out reclamation projects after the mine is closed. Bulldozers, drills, explosives and trucks are all necessary for excavating the land. In the case of placer mining, unconsolidated gravel, or alluvium, is fed into machinery consisting of a hopper and a shaking screen or trommel which frees the desired minerals from the waste gravel. The minerals are then concentrated using sluices or jigs.

The Bagger 288 is a bucket-wheel excavator used in strip mining. It is also the largest land vehicle of all time.

A Bucyrus Erie 2570 dragline and CAT 797 haul truck at the North Antelope Rochelle opencut coal mine

Large drills are used to sink shafts, excavate stopes, and obtain samples for analysis. Trams are used to transport miners, minerals and waste. Lifts carry miners into and out of mines, and move rock and ore out, and machinery in and out, of underground mines. Huge trucks, shovels and cranes are employed in surface mining to move large quantities of overburden and ore. Processing plants utilize large crushers, mills, reactors, roasters and other equipment to consolidate the mineral-rich material and extract the desired compounds and metals from the ore.

Processing

Once the mineral is extracted, it is often then processed. The science of extractive metallurgy is a specialized area in the science of metallurgy that studies the extraction of valuable metals from their ores, especially through chemical or mechanical means.

Mineral processing (or mineral dressing) is a specialized area in the science of metallurgy that studies the mechanical means of crushing, grinding, and washing that enable the separation (extractive metallurgy) of valuable metals or minerals from their gangue (waste material). Processing of placer ore material consists of gravity-dependent methods of separation, such as sluice boxes. Only minor shaking or washing may be necessary to disaggregate (unclump) the sands or gravels before processing. Processing of ore from a lode mine, whether it is a surface or subsurface mine, requires that the rock ore be crushed and pulverized before extraction of the valuable minerals begins. After lode ore is crushed, recovery of the valuable minerals is done by one, or a combination of several, mechanical and chemical techniques.

Since most metals are present in ores as oxides or sulfides, the metal needs to be reduced to its metallic form. This can be accomplished through chemical means such as smelting or through electrolytic reduction, as in the case of aluminium. Geometallurgy combines the geologic sciences with extractive metallurgy and mining.

Environmental Effects

Environmental issues can include erosion, formation of sinkholes, loss of biodiversity, and contamination of soil, groundwater and surface water by chemicals from mining processes. In some cases, additional forest logging is done in the vicinity of mines to create space for the storage of the created debris and soil. Contamination resulting from leakage of chemicals can also affect the health of the local population if not properly controlled. Extreme examples of pollution from mining activities include coal fires, which can last for years or even decades, producing massive amounts of environmental damage.

Iron hydroxide precipitate stains a stream receiving acid drainage from surface coal mining.

Mining companies in most countries are required to follow stringent environmental and rehabilitation codes in order to minimize environmental impact and avoid impacting human health. These codes and regulations all require the common steps of environmental impact assessment, development of environmental management plans, mine closure planning (which must be done before the start of mining operations), and environmental monitoring during operation and after closure. However, in some areas, particularly in the developing world, government regulations may not be well enforced.

For major mining companies and any company seeking international financing, there are a number of other mechanisms to enforce good environmental standards. These generally relate to financing standards such as the Equator Principles, IFC environmental standards, and criteria for Socially responsible investing. Mining companies have used this oversight from the financial sector to argue for some level of industry self-regulation. In 1992, a Draft Code of Conduct for Transnational Corporations was proposed at the Rio Earth Summit by the UN Centre for Transnational Corporations (UNCTC), but the Business Council for Sustainable Development (BCSD) together with the International Chamber of Commerce (ICC) argued successfully for self-regulation instead.

This was followed by the Global Mining Initiative which was begun by nine of the largest metals and mining companies and which led to the formation of the International Council on Mining and Metals, whose purpose was to "act as a catalyst" in an effort to improve social and environmental performance in the mining and metals industry internationally. The mining industry has provided funding to various conservation groups, some of which have been working with conservation agendas that are at odds with an emerging acceptance of the rights of indigenous people – particularly the right to make land-use decisions.

Certification of mines with good practices occurs through the International Organization for Standardization (ISO). For example, ISO 9000 and ISO 14001, which certify an "auditable environmental management system", involve short inspections, although they have been accused of lacking rigor Certification is also available through Ceres' Global Reporting Initiative, but these reports are voluntary and unverified. Miscellaneous other certification programs exist for various projects, typically through nonprofit groups.

The purpose of a 2012 EPS PEAKS paper was to provide evidence on policies managing ecological costs and maximise socio-economic benefits of mining using host country regulatory initiatives. It found existing literature suggesting donors encourage developing countries to:

- Make the environment-poverty link and introduce cutting-edge wealth measures and natural capital accounts.

- Reform old taxes in line with more recent financial innovation, engage directly with the companies, enacting land use and impact assessments, and incorporate specialised support and standards agencies.

- Set in play transparency and community participation initiatives using the wealth accrued.

Waste

Ore mills generate large amounts of waste, called tailings. For example, 99 tons of waste are generated per ton of copper, with even higher ratios in gold mining - because only 5.3 g of gold is extracted per ton of ore, a ton of gold produces 200,000 tons of tailings. These tailings can be toxic. Tailings, which are usually produced as a slurry, are most commonly dumped into ponds made from naturally existing valleys. These ponds are secured by impoundments (dams or embankment dams). In 2000 it was estimated that 3,500 tailings impoundments existed, and that every year, 2 to 5 major failures and 35 minor failures occurred; for example, in the Marcopper mining disaster at least 2 million tons of tailings were released into a local river. Subaqueous tailings disposal is another option. The mining industry has argued that submarine tailings disposal (STD), which disposes of tailings in the sea, is ideal because it avoids the risks of tailings ponds; although the practice is illegal in the United States and Canada, it is used in the developing world.

The waste is classified as either sterile or mineralised, with acid generating potential, and the movement and storage of this material forms a major part of the mine planning process. When the mineralised package is determined by an economic cut-off, the near-grade mineralised waste is usually dumped separately with view to later treatment should market conditions change and it becomes economically viable. Civil engineering design parameters are used in the design of the waste dumps, and special conditions apply to high-rainfall areas and to seismically active areas. Waste dump designs must meet all regulatory requirements of the country in whose jurisdiction the mine is located. It is also common practice to rehabilitate dumps to an internationally acceptable standard, which in some cases means that higher standards than the local regulatory standard are applied.

Renewable Energy and Mining

Many mining sites are remote and not connected to the grid. Electricity is typically generated with diesel generators. Due to high transportation cost and theft during transportation the cost for generating electricity is normally high. Renewable energy applications are becoming an alternative or amendment. Both solar and wind power plants can contribute in saving diesel costs at mining sites. Renewable energy applications have been built at mining sites. Cost savings can reach up to 70%.

Mining Industry

Mining exists in many countries. London is known as the capital of global "mining houses" such as Rio Tinto Group, BHP Billiton, and Anglo American PLC. The US mining industry is also large, but it is dominated by the coal and other nonmetal minerals (e.g., rock and sand), and various regulations have worked to reduce the significance of mining in the United States. In 2007 the total market capitalization of mining companies was reported at US$962 billion, which compares to a total global market cap of publicly traded companies of about US$50 trillion in 2007. In 2002, Chile and Peru were reportedly the major mining countries of South America. The mineral industry of Africa includes the mining of various minerals; it produces relatively little of the industrial metals copper, lead, and zinc, but according to one estimate has as a percent of world reserves 40% of gold, 60% of cobalt, and 90% of the world's platinum group metals. Mining in India is a significant part of that country's economy. In the developed world, mining in Australia, with BHP Billiton founded and headquartered in the country, and mining in Canada are particularly significant. For rare earth minerals mining, China reportedly controlled 95% of production in 2013.

The Bingham Canyon Mine of Rio Tinto's subsidiary, Kennecott Utah Copper.

While exploration and mining can be conducted by individual entrepreneurs or small businesses, most modern-day mines are large enterprises requiring large amounts of capital to establish. Consequently, the mining sector of the industry is dominated by large, often multinational, companies, most of them publicly listed. It can be argued that what is referred to as the 'mining industry' is actually two sectors, one specializing in exploration for new resources and the other in mining those resources. The exploration sector is typically made up of individuals and small mineral resource companies, called "juniors", which are dependent on venture capital. The mining sector is made up of large multinational companies that are sustained by production from their mining operations. Various other industries such as equipment manufacture, environmental testing, and metallurgy analysis rely on, and support, the mining industry throughout the world. Canadian stock exchanges have a particular focus on mining companies, particularly junior exploration companies through Toronto's TSX Venture Exchange; Canadian companies raise capital on these exchanges and then invest the money in exploration globally. Some have argued that below juniors there exists a substantial sector of illegitimate companies primarily focused on manipulating stock prices.

Mining operations can be grouped into five major categories in terms of their respective resources. These are oil and gas extraction, coal mining, metal ore mining, nonmetallic mineral mining and quarrying, and mining support activities. Of all of these categories, oil and gas extraction remains one of the largest in terms of its global economic importance. Prospecting potential mining sites, a vital area of concern for the mining industry, is now done using sophisticated new technologies such as seismic prospecting and remote-sensing satellites. Mining is heavily affected by the prices of the commodity minerals, which are often volatile. The 2000s commodities boom ("commodities supercycle") increased the prices of commodities, driving aggressive mining. In addition, the price of gold increased dramatically in the 2000s, which increased gold mining; for example, one study found that conversion of forest in the Amazon increased six-fold from the period 2003–2006 (292 ha/yr) to the period 2006–2009 (1,915 ha/yr), largely due to artisanal mining.

Corporate Classifications

Mining companies can be classified based on their size and financial capabilities:

- Major companies are considered to have an adjusted annual mining-related revenue of more than US$500 million, with the financial capability to develop a major mine on its own.

- Intermediate companies have at least $50 million in annual revenue but less than $500 million.

- Junior companies rely on equity financing as their principal means of funding exploration. Juniors are mainly pure exploration companies, but may also produce minimally, and do not have a revenue exceeding US$50 million.

Regulation and Governance

New regulation and process of legislative reforms aims to enrich the harmonization and stability of the mining sector in mineral-rich countries. The new legislation for mining industry in the African countries still appears as an emerging issue with a potential to be solved, until a consensus is reached on the best approach. By the beginning of the 21st century the booming and more complex mining sector in mineral-rich countries provided only slight benefits to local communities in terms of sustainability. Increasing debates and influence by NGOs and communities appealed for a new program which would have had also included a disadvantaged communities, and would have had worked towards sustainable development even after mine closure (included transparency and revenue management). By the early 2000s, community development issues and resettlements became mainstreamed in World Bank mining projects. Mining-industry expansion after an increase of mineral prices in 2003 and also potential fiscal revenues in those countries created an omission in the other economic sectors in terms of finances and development. Furthermore, it had highlighted regional and local demand of mining-revenues and lack of ability of sub-national governments to use the revenues. The Fraser Institute (a Canadian think tank) has highlighted the environmental protection laws in developing countries, as well as the voluntary efforts by mining companies to improve their environmental impact.

In 2007 the Extractive Industries Transparency Initiative (EITI) was mainstreamed in all countries cooperating with the World Bank in mining industry reform. The EITI is operating and implementing with a support of EITI Multi-Donor Trust Fund, managed by The World Bank. The Extractive Industries Transparency Initiative (EITI) aims to increase transparency in transactions between governments and companies within extractive industries by monitoring the revenues and benefits between industries and recipient governments. The entrance process is voluntary for each country and is being monitored by multi-stakeholders involving government, private companies and civil society representatives, responsible for disclosure and dissemination of the reconciliation report; however, the competitive disadvantage of company-by company public report is for some of the businesses in Ghana, the main constraint. Therefore, the outcome assessment in terms of failure or success of the new EITI regulation does not only "rest on the government's shoulders" but also on civil society and companies.

On the other hand, criticism points out two main implementation issues; inclusion or exclusion of artisanal mining and small-scale mining (ASM) from the EITI and how to deal with "non-cash" payments made by companies to subnational governments. Furthermore, disproportion of the revenues mining industry creates to the comparatively small number of people that it employs, causes another controversy. The issue of artisanal mining is clearly an issue in EITI Countries such as the Central African Republic, D.R. Congo, Guinea, Liberia and Sierra Leone – i.e. almost half of the mining countries implementing the EITI. Among other things, limited scope of the EITI involving disparity in terms of knowledge of the industry and negotiation skills, thus far flexibility of the policy (e.g. liberty of the countries to expand beyond the minimum requirements and adapt it to their needs), creates another risk of unsuccessful implementation. Public awareness increase, where government should act as a bridge between public and initiative for a successful outcome of the policy is an important element to be considered.

World Bank

The World Bank has been involved in mining since 1955, mainly through grants from its International Bank for Reconstruction and Development, with the Bank's Multilateral Investment Guarantee Agency offering political risk insurance. Between 1955 and 1990 it provided about $2 billion to fifty mining projects, broadly categorized as reform and rehabilitation, greenfield mine construction, mineral processing, technical assistance, and engineering. These projects have been criticized, particularly the Ferro Carajas project of Brazil, begun in 1981. The World Bank established mining codes intended to increase foreign investment; in 1988 it solicited feedback from 45 mining companies on how to increase their involvement.

In 1992 the World Bank began to push for privatization of government-owned mining companies with a new set of codes, beginning with its report *The Strategy for African*

Mining. In 1997, Latin America's largest miner Companhia Vale do Rio Doce (CVRD) was privatized. These and other developments such as the Philippines 1995 Mining Act led the bank to publish a third report (*Assistance for Minerals Sector Development and Reform in Member Countries*) which endorsed mandatory environment impact assessments and attention to the concerns of the local population. The codes based on this report are influential in the legislation of developing nations. The new codes are intended to encourage development through tax holidays, zero custom duties, reduced income taxes, and related measures. The results of these codes were analyzed by a group from the University of Quebec, which concluded that the codes promote foreign investment but "fall very short of permitting sustainable development". The observed negative correlation between natural resources and economic development is known as the resource curse.

Safety

Safety has long been a concern in the mining business, especially in sub-surface mining. The Courrières mine disaster, Europe's worst mining accident, involved the death of 1,099 miners in Northern France on March 10, 1906. This disaster was surpassed only by the Benxihu Colliery accident in China on April 26, 1942, which killed 1,549 miners. While mining today is substantially safer than it was in previous decades, mining accidents still occur. Government figures indicate that 5,000 Chinese miners die in accidents each year, while other reports have suggested a figure as high as 20,000. Mining accidents continue worldwide, including accidents causing dozens of fatalities at a time such as the 2007 Ulyanovskaya Mine disaster in Russia, the 2009 Heilongjiang mine explosion in China, and the 2010 Upper Big Branch Mine disaster in the United States.

There are numerous occupational hazards associated with mining, including exposure to rockdust which can lead to diseases such as silicosis, asbestosis, and pneumoconiosis. Gases in the mine can lead to asphyxiation and could also be ignited. Mining equipment can generate considerable noise, putting workers at risk for hearing loss. Cave-ins, rock falls, and exposure to excess heat are also known hazards.

Proper ventilation, hearing protection, and spraying equipment with water are important safety practices in mines.

Records

As of 2008, the deepest mine in the world is TauTona in Carletonville, South Africa at 3.9 kilometres (2.4 mi), replacing the neighboring Savuka Mine in the North West Province of South Africa at 3,774 metres (12,382 ft). East Rand Mine in Boksburg, South Africa briefly held the record at 3,585 metres (11,762 ft), and the first mine declared the deepest in the world was also TauTona when it was at 3,581 metres (11,749 ft).

Chuquicamata, Chile, site of the largest circumference and second deepest
open pit copper mine in the world.

The Moab Khutsong gold mine in North West Province (South Africa) has the world's longest winding steel wire rope, able to lower workers to 3,054 metres (10,020 ft) in one uninterrupted four-minute journey.

The deepest mine in Europe is the 16th shaft of the uranium mines in Příbram, Czech Republic at 1,838 metres (6,030 ft), second is Bergwerk Saar in Saarland, Germany at 1,750 metres (5,740 ft).

The deepest open-pit mine in the world is Bingham Canyon Mine in Bingham Canyon, Utah, United States at over 1,200 metres (3,900 ft). The largest and second deepest open-pit copper mine in the world is Chuquicamata in Chuquicamata, Chile at 900 metres (3,000 ft), 443,000 tons of copper and 20,000 tons of molybdenum produced annually.

The deepest open-pit mine with respect to sea level is Tagebau Hambach in Germany, where the base of the pit is 293 metres (961 ft) below sea level.

The largest underground mine is Kiirunavaara Mine in Kiruna, Sweden. With 450 kilometres (280 mi) of roads, 40 million tonnes of ore produced yearly, and a depth of 1,270 metres (4,170 ft), it is also one of the most modern underground mines. The deepest borehole in the world is Kola Superdeep Borehole at 12,262 metres (40,230 ft). This, however, is not a matter of mining but rather related to scientific drilling.

Metal Reserves and Recycling

During the twentieth century, the variety of metals used in society grew rapidly. Today, the development of major nations such as China and India and advances in technologies are fueling an ever greater demand. The result is that metal mining activities are expanding and more and more of the world's metal stocks are above ground in use rather than below ground as unused reserves. An example is the in-use stock of copper. Between 1932 and 1999, copper in use in the USA rose from 73 kilograms (161 lb) to 238 kilograms (525 lb) per person.

95% of the energy used to make aluminium from bauxite ore is saved by using recycled material. However, levels of metals recycling are generally low. In 2010, the International Resource Panel, hosted by the United Nations Environment Programme (UNEP), published reports on metal stocks that exist within society and their recycling rates.

The report's authors observed that the metal stocks in society can serve as huge mines above ground. However, they warned that the recycling rates of some rare metals used in applications such as mobile phones, battery packs for hybrid cars, and fuel cells are so low that unless future end-of-life recycling rates are dramatically stepped up these critical metals will become unavailable for use in modern technology.

As recycling rates are low and so much metal has already been extracted, some landfills now contain higher concentrations of metal than mines themselves. This is especially true with aluminium, found in cans, and precious metals in discarded electronics. Furthermore, waste after 15 years has still not broken down, so less processing would be required when compared to mining ores. A study undertaken by Cranfield University has found £360 million of metals could be mined from just 4 landfill sites. There is also up to 20MW/kg of energy in waste, potentially making the re-extraction more profitable. However, although the first landfill mine opened in Tel Aviv, Israel in 1953, little work has followed due to the abundance of accessible ores.

Geotechnical Engineering

Geotechnical engineering is the branch of civil engineering concerned with the engineering behavior of earth materials. Geotechnical engineering is important in civil engineering, but also has applications in military, mining, petroleum and other engineering disciplines that are concerned with construction occurring on the surface or within the ground. Geotechnical engineering uses principles of soil mechanics and rock mechanics to investigate subsurface conditions and materials; determine the relevant physical/mechanical and chemical properties of these materials; evaluate stability of natural slopes and man-made soil deposits; assess risks posed by site conditions; design earthworks and structure foundations; and monitor site conditions, earthwork and foundation construction.

A typical geotechnical engineering project begins with a review of project needs to define the required material properties. Then follows a site investigation of soil, rock, fault distribution and bedrock properties on and below an area of interest to determine their engineering properties including how they will interact with, on or in a proposed construction. Site investigations are needed to gain an understanding of the area in or on which the engineering will take place. Investigations can include the assessment of the risk to humans, property and the environment from natural hazards such as earthquakes, landslides, sinkholes, soil liquefaction, debris flows and rockfalls.

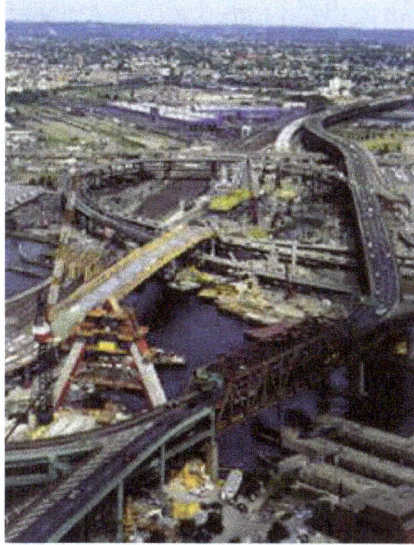

Boston's Big Dig presented geotechnical challenges in an urban environment.

A geotechnical engineer then determines and designs the type of foundations, earth-works, and/or pavement subgrades required for the intended man-made structures to be built. Foundations are designed and constructed for structures of various sizes such as high-rise buildings, bridges, medium to large commercial buildings, and smaller structures where the soil conditions do not allow code-based design.

Foundations built for above-ground structures include shallow and deep foundations. Retaining structures include earth-filled dams and retaining walls. Earthworks include embankments, tunnels, dikes and levees, channels, reservoirs, deposition of hazardous waste and sanitary landfills.

Geotechnical engineering is also related to coastal and ocean engineering. Coastal engineering can involve the design and construction of wharves, marinas, and jetties. Ocean engineering can involve foundation and anchor systems for offshore structures such as oil platforms.

The fields of geotechnical engineering and engineering geology are closely related, and have large areas of overlap. However, the field of geotechnical engineering is a specialty of engineering, where the field of engineering geology is a specialty of geology.

History

Humans have historically used soil as a material for flood control, irrigation purposes, burial sites, building foundations, and as construction material for buildings. First activities were linked to irrigation and flood control, as demonstrated by traces of dykes, dams, and canals dating back to at least 2000 BCE that were found in ancient Egypt, ancient Mesopotamia and the Fertile Crescent, as well as around the early settlements of Mohenjo Daro and Harappa in the Indus valley. As the cities expanded, structures

were erected supported by formalized foundations; Ancient Greeks notably constructed pad footings and strip-and-raft foundations. Until the 18th century, however, no
theoretical basis for soil design had been developed and the discipline was more of an
art than a science, relying on past experience.

Several foundation-related engineering problems, such as the Leaning Tower of Pisa,
prompted scientists to begin taking a more scientific-based approach to examining
the subsurface. The earliest advances occurred in the development of earth pressure
theories for the construction of retaining walls. Henri Gautier, a French Royal Engineer, recognized the "natural slope" of different soils in 1717, an idea later known as
the soil's angle of repose. A rudimentary soil classification system was also developed
based on a material's unit weight, which is no longer considered a good indication of
soil type.

The application of the principles of mechanics to soils was documented as early as
1773 when Charles Coulomb (a physicist, engineer, and army Captain) developed improved methods to determine the earth pressures against military ramparts. Coulomb
observed that, at failure, a distinct slip plane would form behind a sliding retaining
wall and he suggested that the maximum shear stress on the slip plane, for design
purposes, was the sum of the soil cohesion, , and friction , where is the normal stress
on the slip plane and is the friction angle of the soil. By combining Coulomb's theory
with Christian Otto Mohr's 2D stress state, the theory became known as Mohr-Coulomb theory. Although it is now recognized that precise determination of cohesion is
impossible because is not a fundamental soil property, the Mohr-Coulomb theory is
still used in practice today.

In the 19th century Henry Darcy developed what is now known as Darcy's Law describing the flow of fluids in porous media. Joseph Boussinesq (a mathematician and
physicist) developed theories of stress distribution in elastic solids that proved useful
for estimating stresses at depth in the ground; William Rankine, an engineer and physicist, developed an alternative to Coulomb's earth pressure theory. Albert Atterberg
developed the clay consistency indices that are still used today for soil classification.
Osborne Reynolds recognized in 1885 that shearing causes volumetric dilation of dense
and contraction of loose granular materials.

Modern geotechnical engineering is said to have begun in 1925 with the publication
of *Erdbaumechanik* by Karl Terzaghi (a civil engineer and geologist). Considered by
many to be the father of modern soil mechanics and geotechnical engineering, Terzaghi developed the principle of effective stress, and demonstrated that the shear
strength of soil is controlled by effective stress. Terzaghi also developed the framework for theories of bearing capacity of foundations, and the theory for prediction of
the rate of settlement of clay layers due to consolidation. In his 1948 book, Donald
Taylor recognized that interlocking and dilation of densely packed particles contributed to the peak strength of a soil. The interrelationships between volume change

behavior (dilation, contraction, and consolidation) and shearing behavior were all connected via the theory of plasticity using critical state soil mechanics by Roscoe, Schofield, and Wroth with the publication of "On the Yielding of Soils" in 1958. Critical state soil mechanics is the basis for many contemporary advanced constitutive models describing the behavior of soil.

Geotechnical centrifuge modeling is a method of testing physical scale models of geotechnical problems. The use of a centrifuge enhances the similarity of the scale model tests involving soil because the strength and stiffness of soil is very sensitive to the confining pressure. The centrifugal acceleration allows a researcher to obtain large (prototype-scale) stresses in small physical models.

Practicing Engineers

Geotechnical engineers are typically graduates of a four-year civil engineering program and some hold a masters degree. In the USA, geotechnical engineers are typically licensed and regulated as Professional Engineers (PEs) in most states; currently only California and Oregon have licensed geotechnical engineering specialties. The Academy of Geo-Professionals (AGP) began issuing Diplomate, Geotechnical Engineering (D.GE) certification in 2008. State governments will typically license engineers who have graduated from an ABET accredited school, passed the Fundamentals of Engineering examination, completed several years of work experience under the supervision of a licensed Professional Engineer, and passed the Professional Engineering examination.

Soil Mechanics

In geotechnical engineering, soils are considered a three-phase material composed of: rock or mineral particles, water and air. The voids of a soil, the spaces in between mineral particles, contain the water and air.

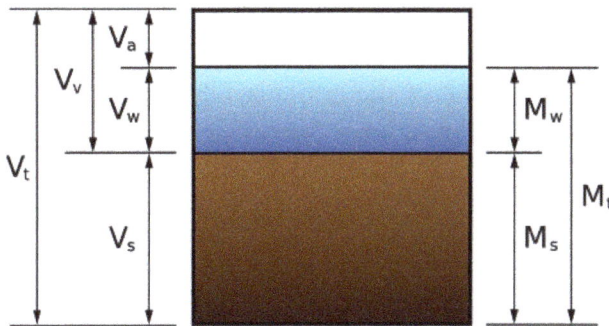

A phase diagram of soil indicating the weights and volumes of air, soil, water, and voids.

The engineering properties of soils are affected by four main factors: the predominant size of the mineral particles, the type of mineral particles, the grain size distribution, and the relative quantities of mineral, water and air present in the soil matrix. Fine particles (fines) are defined as particles less than 0.075 mm in diameter.

Soil Properties

Some of the important properties of soils that are used by geotechnical engineers to analyze site conditions and design earthworks, retaining structures, and foundations are:

Specific weight or Unit Weight

Cumulative weight of the solid particles, water and air of the unit volume of soil. Note that the air phase is often assumed to be weightless.

Porosity

Ratio of the volume of voids (containing air, water, or other fluids) in a soil to the total volume of the soil. Porosity is mathematically related to void ratio the by

here e is void ratio and n is porosity

Void ratio

The ratio of the volume of voids to the volume of solid particles in a soil mass. Void ratio is mathematically related to the porosity by

Permeability

A measure of the ability of water to flow through the soil. It is expressed in units of velocity.

Compressibility

The rate of change of volume with effective stress. If the pores are filled with water, then the water must be squeezed out of the pores to allow volumetric compression of the soil; this process is called consolidation.

Shear strength

The maximum shear stress that can be applied in a soil mass without causing shear failure.

Atterberg Limits

Liquid limit, Plastic limit, and Shrinkage limit. These indices are used for estimation of other engineering properties and for soil classification.

Geotechnical Investigation

Geotechnical engineers and engineering geologists perform geotechnical investigations to obtain information on the physical properties of soil and rock underlying (and sometimes adjacent to) a site to design earthworks and foundations for proposed structures, and for repair of distress to earthworks and structures caused by subsurface conditions. A geotechnical investigation will include surface exploration and subsurface exploration of a site. Sometimes, geophysical methods are used to obtain data about sites. Subsurface exploration usually involves in-situ testing (two common examples of in-situ tests are the standard penetration test and cone penetration test). In addition site investigation will often include subsurface sampling and laboratory testing of the soil samples retrieved. The digging of test pits and trenching (particularly for locating faults and slide planes) may also be used to learn about soil conditions at depth. Large diameter borings are rarely used due to safety concerns and expense, but are sometimes used to allow a geologist or engineer to be lowered into the borehole for direct visual and manual examination of the soil and rock stratigraphy.

A variety of soil samplers exist to meet the needs of different engineering projects. The standard penetration test (SPT), which uses a thick-walled split spoon sampler, is the most common way to collect disturbed samples. Piston samplers, employing a thin-walled tube, are most commonly used for the collection of less disturbed samples. More advanced methods, such as ground freezing and the Sherbrooke block sampler, are superior, but even more expensive.

Atterberg limits tests, water content measurements, and grain size analysis, for example, may be performed on disturbed samples obtained from thick walled soil samplers. Properties such as shear strength, stiffness hydraulic conductivity, and coefficient of consolidation may be significantly altered by sample disturbance. To measure these properties in the laboratory, high quality sampling is required. Common tests to measure the strength and stiffness include the triaxial shear and unconfined compression test.

Surface exploration can include geologic mapping, geophysical methods, and photogrammetry; or it can be as simple as an engineer walking around to observe the physical conditions at the site. Geologic mapping and interpretation of geomorphology is typically completed in consultation with a geologist or engineering geologist.

Geophysical exploration is also sometimes used. Geophysical techniques used for subsurface exploration include measurement of seismic waves (pressure, shear, and Rayleigh waves), surface-wave methods and/or downhole methods, and electromagnetic surveys (magnetometer, resistivity, and ground-penetrating radar).

Foundations

A building's foundation transmits loads from buildings and other structures to the earth. Geotechnical engineers design foundations based on the load characteristics of the structure and the properties of the soils and/or bedrock at the site. In general, geotechnical engineers:

1. Estimate the magnitude and location of the loads to be supported.

2. Develop an investigation plan to explore the subsurface.

3. Determine necessary soil parameters through field and lab testing (e.g., consolidation test, triaxial shear test, vane shear test, standard penetration test).

4. Design the foundation in the safest and most economical manner.

The primary considerations for foundation support are bearing capacity, settlement, and ground movement beneath the foundations. Bearing capacity is the ability of the site soils to support the loads imposed by buildings or structures. Settlement occurs under all foundations in all soil conditions, though lightly loaded structures or rock sites may experience negligible settlements. For heavier structures or softer sites, both overall settlement relative to unbuilt areas or neighboring buildings, and differential settlement under a single structure, can be concerns. Of particular concern is settlement which occurs over time, as immediate settlement can usually be compensated for during construction. Ground movement beneath a structure's foundations can occur due to shrinkage or swell of expansive soils due to climatic changes, frost expansion of soil, melting of permafrost, slope instability, or other causes. All these factors must be considered during design of foundations.

Many building codes specify basic foundation design parameters for simple conditions, frequently varying by jurisdiction, but such design techniques are normally limited to certain types of construction and certain types of sites, and are frequently very conservative.

In areas of shallow bedrock, most foundations may bear directly on bedrock; in other areas, the soil may provide sufficient strength for the support of structures. In areas of deeper bedrock with soft overlying soils, deep foundations are used to support structures directly on the bedrock; in areas where bedrock is not economically available, stiff "bearing layers" are used to support deep foundations instead.

Shallow Foundations

Example of a slab-on-grade foundation.

Shallow foundations are a type of foundation that transfers building load to the very near the surface, rather than to a subsurface layer. Shallow foundations typically have a depth to width ratio of less than 1.

Footings

Footings (often called "spread footings" because they spread the load) are structural elements which transfer structure loads to the ground by direct areal contact. Footings can be isolated footings for point or column loads, or strip footings for wall or other long (line) loads. Footings are normally constructed from reinforced concrete cast directly onto the soil, and are typically embedded into the ground to penetrate through the zone of frost movement and/or to obtain additional bearing capacity.

Slab Foundations

A variant on spread footings is to have the entire structure bear on a single slab of concrete underlying the entire area of the structure. Slabs must be thick enough to provide sufficient rigidity to spread the bearing loads somewhat uniformly, and to minimize differential settlement across the foundation. In some cases, flexure is allowed and the building is constructed to tolerate small movements of the foundation instead. For small structures, like single-family houses, the slab may be less than 300 mm thick; for larger structures, the foundation slab may be several meters thick.

Slab foundations can be either slab-on-grade foundations or embedded foundations, typically in buildings with basements. Slab-on-grade foundations must be designed to allow for potential ground movement due to changing soil conditions.

Deep Foundations

Pile-driving for a bridge in Napa, California.

Deep foundations are used for structures or heavy loads when shallow foundations cannot provide adequate capacity, due to size and structural limitations. They may also be used to transfer building loads past weak or compressible soil layers. While shallow foundations rely solely on the bearing capacity of the soil beneath them, deep foundations can rely on end bearing resistance, frictional resistance along their length, or both in developing the required capacity. Geotechnical engineers use specialized tools, such as the cone penetration test, to estimate the amount of skin and end bearing resistance available in the subsurface.

There are many types of deep foundations including piles, drilled shafts, caissons, piers, and earth stabilized columns. Large buildings such as skyscrapers typically require deep foundations. For example, the Jin Mao Tower in China uses tubular steel piles about 1m (3.3 feet) driven to a depth of 83.5m (274 feet) to support its weight.

In buildings that are constructed and found to undergo settlement, underpinning piles can be used to stabilise the existing building.

There are three ways to place piles for a deep foundation. They can be driven, drilled, or installed by use of an auger. Driven piles are extended to their necessary depths with the application of external energy in the same way a nail is hammered. There are four typical hammers used to drive such piles: drop hammers, diesel hammers, hydraulic hammers, and air hammers. Drop hammers simply drop a heavy weight onto the pile to drive it, while diesel hammers use a single cylinder diesel engine to force piles through the Earth. Similarly, hydraulic and air hammers supply energy to piles through hydraulic and air forces. Energy imparted from a hammer head varies with type of hammer chosen, and can be as high as a million foot pounds for large scale diesel hammers, a very common hammer head used in practice. Piles are made of a variety of material including steel, timber, and concrete. Drilled piles are created by first drilling a hole to the appropriate depth, and filling it with concrete. Drilled piles can typically carry more load than driven piles, simply due to a larger diameter pile. The auger method of pile installation is similar to drilled pile installation, but concrete is pumped into the hole as the auger is being removed.

Lateral Earth Support Structures

A retaining wall is a structure that holds back earth. Retaining walls stabilize soil and rock from downslope movement or erosion and provide support for vertical or near-vertical grade changes. Cofferdams and bulkheads, structures to hold back water, are sometimes also considered retaining walls.

The primary geotechnical concern in design and installation of retaining walls is that the weight of the retained material is creates lateral earth pressure behind the wall, which can cause the wall to deform or fail. The lateral earth pressure depends on the height of the wall, the density of the soil,the strength of the soil, and the amount of allowable movement of the wall. This pressure is smallest at the top and increases toward the bottom in a manner similar to hydraulic pressure, and tends to push the wall away from the backfill. Groundwater behind the wall that is not dissipated by a drainage system causes an additional horizontal hydraulic pressure on the wall.

Gravity Walls

Gravity walls depend on the size and weight of the wall mass to resist pressures from behind. Gravity walls will often have a slight setback, or batter, to improve wall stability. For short, landscaping walls, gravity walls made from dry-stacked (mortarless) stone or segmental concrete units (masonry units) are commonly used.

Earlier in the 20th century, taller retaining walls were often gravity walls made from large masses of concrete or stone. Today, taller retaining walls are increasingly built as composite gravity walls such as: geosynthetic or steel-reinforced backfill soil with precast facing; gabions (stacked steel wire baskets filled with rocks), crib walls (cells built up log cabin style from precast concrete or timber and filled with soil or free draining gravel) or soil-nailed walls (soil reinforced in place with steel and concrete rods).

For reinforced-soil gravity walls, the soil reinforcement is placed in horizontal layers throughout the height of the wall. Commonly, the soil reinforcement is geogrid, a high-strength polymer mesh, that provide tensile strength to hold soil together. The wall face is often of precast, segmental concrete units that can tolerate some differential movement. The reinforced soil's mass, along with the facing, becomes the gravity wall. The reinforced mass must be built large enough to retain the pressures from the soil behind it. Gravity walls usually must be a minimum of 30 to 40 percent as deep (thick) as the height of the wall, and may have to be larger if there is a slope or surcharge on the wall.

Cantilever Walls

Prior to the introduction of modern reinforced-soil gravity walls, cantilevered walls were the most common type of taller retaining wall. Cantilevered walls are made from a relatively thin stem of steel-reinforced, cast-in-place concrete or mortared masonry (often in the shape of an inverted T). These walls cantilever loads (like a beam) to a large, structural footing; converting horizontal pressures from behind the wall to vertical pressures on the ground below. Sometimes cantilevered walls are buttressed on the front, or include a counterfort on the back, to improve their stability against high loads. Buttresses are short wing walls at right angles to the main trend of the wall. These walls require rigid concrete footings below seasonal frost depth. This type of wall uses much less material than a traditional gravity wall.

Cantilever walls resist lateral pressures by friction at the base of the wall and/or passive earth pressure, the tendency of the soil to resist lateral movement.

Basements are a form of cantilever walls, but the forces on the basement walls are greater than on conventional walls because the basement wall is not free to move.

Excavation Shoring

Shoring of temporary excavations frequently requires a wall design which does not extend laterally beyond the wall, so shoring extends below the planned base of the excavation. Common methods of shoring are the use of sheet piles or soldier beams and lagging. Sheet piles are a form of driven piling using thin interlocking sheets of steel to obtain a continuous barrier in the ground, and are driven prior to excavation. Soldier beams are constructed of wide flange steel H sections spaced about 2–3 m apart, driven prior to excavation. As the excavation proceeds, horizontal timber or steel sheeting (lagging) is inserted behind the H pile flanges.

In some cases, the lateral support which can be provided by the shoring wall alone is insufficient to resist the planned lateral loads; in this case additional support is provided by walers or tie-backs. Walers are structural elements which connect across the excavation so that the loads from the soil on either side of the excavation are used to resist each other, or which transfer horizontal loads from the shoring wall to the base of the excavation. Tie-backs are steel tendons drilled into the face of the wall which extend beyond the soil which is applying pressure to the wall, to provide additional lateral resistance to the wall.

Earthworks

Excavation

Excavation is the process of training earth according to requirement by removing the soil from the site.

Filling

Filling is the process of training earth according to requirement by placing the soil on the site.

Compaction

A compactor/roller operated by U.S. Navy Seabees

Compaction is the process by which the density of soil is increased and permeability of soil is decreased. Fill placement work often has specifications requiring a specific degree of compaction, or alternatively, specific properties of the compacted soil. In-situ soils can be compacted by rolling, deep dynamic compaction, vibration, blasting, gyrating, kneading, compaction grouting etc.

Ground Improvement

Ground Improvement is a technique that improves the engineering properties of the treated soil mass. Usually, the properties modified are shear strength, stiffness and permeability. Ground improvement has developed into a sophisticated tool to support foundations for a wide variety of structures. Properly applied, i.e. after giving due consideration to the nature of the ground being improved and the type and sensitivity of the structures being built, ground improvement often reduces direct costs and saves time.

Slope Stabilization

Slope stability is the potential of soil covered slopes to withstand and undergo movement. Stability is determined by the balance of shear stress and shear strength. A previously stable slope may be initially affected by preparatory factors, making the slope conditionally unstable. Triggering factors of a slope failure can be climatic events can then make a slope actively unstable, leading to mass movements. Mass movements can be caused by increases in shear stress, such as loading, lateral pressure, and transient forces. Alternatively, shear strength may be decreased by weathering, changes in pore water pressure, and organic material.

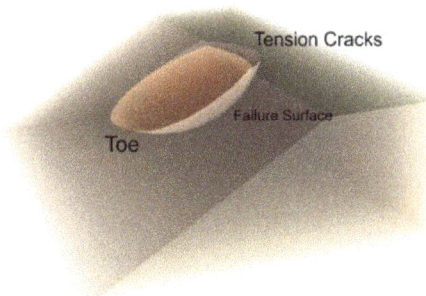

Simple slope slip section.

Several modes of failure for earth slopes include falls, topples, slides, and flows. In slopes with coarse grained soil or rocks, falls typically occur as the rapid descent of rocks and other loose slope material. A slope topples when a large column of soil tilts over its vertical axis at failure. Typical slope stability analysis considers sliding failures, categorized mainly as rotational slides or translational slides. As implied by the name, rotational slides fail along a generally curved surface, while translational slides fail along a more planar surface. A slope failing as a flow would resemble a fluid flowing downhill.

Slope Stability Analysis

Stability analysis is needed for the design of engineered slopes and for estimating the risk of slope failure in natural or designed slopes. A common assumption is that a slope consists of a layer of soil sitting on top of a rigid base. The mass and the base are assumed to interact via friction. The interface between the mass and the base can be planar, curved, or have some other complex geometry. The goal of a slope stability analysis is to determine the conditions under which the mass will slip relative to the base and lead to slope failure.

If the interface between the mass and the base of a slope has a complex geometry, slope stability analysis is difficult and numerical solution methods are required. Typically, the exact geometry of the interface is not known and a simplified interface geometry is assumed. Finite slopes require three-dimensional models to be analyzed. To keep the problem simple, most slopes are analyzed assuming that the slopes are infinitely wide and can therefore be represented by two-dimensional models. A slope can be drained or undrained. The undrained condition is used in the calculations to produce conservative estimates of risk.

A popular stability analysis approach is based on principles pertaining to the limit equilibrium concept. This method analyzes a finite or infinite slope as if it were about to fail along its sliding failure surface. Equilibrium stresses are calculated along the failure plane, and compared to the soils shear strength as determined by Terzaghi's shear strength equation. Stability is ultimately decided by a factor of safety equal to the ratio of shear strength to the equilibrium stresses along the failure surface. A factor of safety greater than one generally implies a stable slope, failure of which should not occur assuming the slope is undisturbed. A factor of safety of 1.5 for static conditions is commonly used in practice.

Offshore Geotechnical Engineering

Platforms offshore Mexico.

Offshore (or *marine*) *geotechnical engineering* is concerned with foundation design for human-made structures in the sea, away from the coastline (in opposition to *onshore* or *nearshore*). Oil platforms, artificial islands and submarine pipelines are examples of such structures. There are number of significant differences between onshore and off-shore geotechnical engineering. Notably, ground improvement (on the seabed) and site investigation are more expensive, the offshore structures are exposed to a wider range of geohazards, and the environmental and financial consequences are higher in case of failure. Offshore structures are exposed to various environmental loads, notably wind, waves and currents. These phenomena may affect the integrity or the serviceability of the structure and its foundation during its operational lifespan – they need to be taken into account in offshore design.

In subsea geotechnical engineering, seabed materials are considered a two-phase material composed of 1) rock or mineral particles and 2) water. Structures may be fixed in place in the seabed—as is the case for piers, jettys and fixed-bottom wind turbines—or may be a floating structure that remain roughly fixed relative to its geotechnical anchor point. Undersea mooring of human-engineered floating structures include a large number of offshore oil and gas platforms and, since 2008, a few floating wind turbines. Two common types of engineered design for anchoring floating structures include tension-leg and catenary loose mooring systems. "Tension leg mooring systems have vertical tethers under tension providing large restoring moments in pitch and roll. Catenary mooring systems provide station keeping for an offshore structure yet provide little stiffness at low tensions."

Geosynthetics

Geosynthetics are a type of plastic polymer products used in geotechnical engineering that improve engineering performance while reducing costs. This includes geotextiles, geogrids, geomembranes, geocells, and geocomposites. The synthetic nature of the products make them suitable for use in the ground where high levels of durability are required; their main functions include: drainage, filtration, reinforcement, separation and containment. Geosynthetics are available in a wide range of forms and materials, each to suit a slightly different end use, although they are frequently used together. These products have a wide range of applications and are currently used in many civil and geotechnical engineering applications including: roads, airfields, railroads, embankments, piled embankments, retaining structures, reservoirs, canals, dams, landfills, bank protection and coastal engineering.

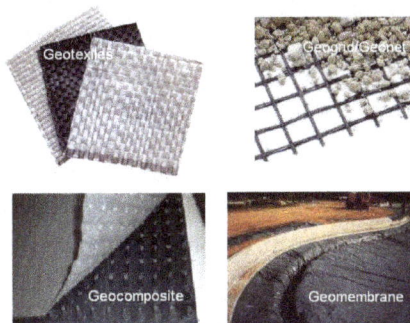

A collage of geosynthetic products.

Plate Tectonics

Plate tectonics (from the Late Latin *tectonicus*) is a scientific theory describing the large-scale motion of Earth's lith-osphere. The theoretical model builds on the concept of continental drift developed during the first few decades of the 20th century. The geoscientific community accepted plate-tectonic theory after seafloor spreading was validated in the late 1950s and early 1960s.

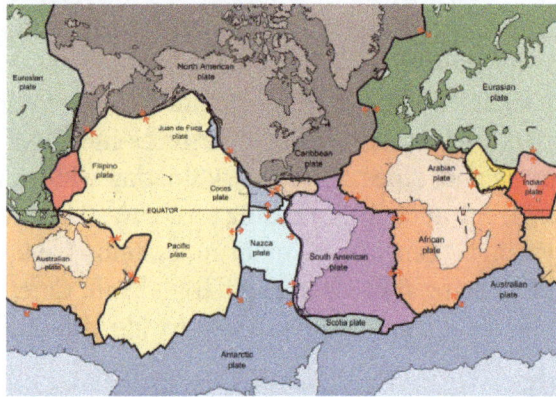

The tectonic plates of the world were mapped in the second half of the 20th century.

The lithosphere, which is the rigid outermost shell of a planet (the crust and upper mantle), is broken up into tectonic plates. The Earth's lithosphere is composed of seven or eight major plates (depending on how they are defined) and many minor plates. Where the plates meet, their relative motion determines the type of boundary: convergent, divergent, or transform. Earthquakes, volcanic activity, mountain-building, and oceanic trench formation occur along these plate boundaries. The relative movement of the plates typically ranges from zero to 100 mm annually.

Tectonic plates are composed of oceanic lithosphere and thicker continental lithosphere, each topped by its own kind of crust. Along convergent boundaries, subduction carries plates into the mantle; the material lost is roughly balanced by the formation of new (oceanic) crust along divergent margins by seafloor spreading. In this way, the total surface of the lithosphere remains the same. This prediction of plate tectonics is also referred to as the conveyor belt principle. Earlier theories (that still have some supporters) propose gradual shrinking (contraction) or gradual expansion of the globe.

Tectonic plates are able to move because the Earth's lithosphere has greater strength than the underlying asthenosphere. Lateral density variations in the mantle result in convection. Plate movement is thought to be driven by a combination of the motion of the seafloor away from the spreading ridge (due to variations in topography and density of the crust, which result in differences in gravitational forces) and drag, with downward suction, at the subduction zones. Another explanation lies in the different forces generated by tidal forces of the Sun and Moon. The relative importance of each of these factors and their relationship to each other is unclear, and still the subject of much debate.

Key Principles

The outer layers of the Earth are divided into the lithosphere and asthenosphere. This is based on differences in mechanical properties and in the method for the transfer of heat. Mechanically, the lithosphere is cooler and more rigid, while the asthenosphere is hotter and flows more easily. In terms of heat transfer, the lithosphere loses heat by conduction, whereas the asthenosphere also transfers heat by convection and has a nearly adiabatic temperature gradient. This division should not be confused with the *chemical* subdivision of these same layers into the mantle (comprising both the asthenosphere and the mantle portion of the lithosphere) and the crust: a given piece of mantle may be part of the lithosphere or the asthenosphere at different times depending on its temperature and pressure.

The key principle of plate tectonics is that the lithosphere exists as separate and distinct *tectonic plates*, which ride on the fluid-like (visco-elastic solid) asthenosphere. Plate motions range up to a typical 10–40 mm/year (Mid-Atlantic Ridge; about as fast as fingernails grow), to about 160 mm/year (Nazca Plate; about as fast as hair grows). The driving mechanism behind this movement is described below.

Tectonic lithosphere plates consist of lithospheric mantle overlain by either or both of two types of crustal material: oceanic crust (in older texts called *sima* from silicon and magnesium) and continental crust (*sial* from silicon and aluminium). Average oceanic lithosphere is typically 100 km (62 mi) thick; its thickness is a function of its age: as time passes, it conductively cools and subjacent cooling mantle is added to its base. Because it is formed at mid-ocean ridges and spreads outwards, its thickness is therefore a function of its distance from the mid-ocean ridge where it was formed. For a typical distance that oceanic lithosphere must travel before being subducted, the thickness varies from about 6 km (4 mi) thick at mid-ocean ridges to greater than 100 km (62 mi) at subduction zones; for shorter or longer distances, the subduction zone (and therefore also the mean) thickness becomes smaller or larger, respectively. Continental lithosphere is typically ~200 km thick, though this varies considerably between basins, mountain ranges, and stable cratonic interiors of continents. The two types of crust also differ in thickness, with continental crust being considerably thicker than oceanic (35 km vs. 6 km).

The location where two plates meet is called a *plate boundary*. Plate boundaries are commonly associated with geological events such as earthquakes and the creation of topographic features such as mountains, volcanoes, mid-ocean ridges, and oceanic trenches. The majority of the world's active volcanoes occur along plate boundaries, with the Pacific Plate's Ring of Fire being the most active and widely known today.

These boundaries are discussed in further detail below. Some volcanoes occur in the interiors of plates, and these have been variously attributed to internal plate deformation and to mantle plumes.

As explained above, tectonic plates may include continental crust or oceanic crust, and most plates contain both. For example, the African Plate includes the continent and parts of the floor of the Atlantic and Indian Oceans. The distinction between oceanic crust and continental crust is based on their modes of formation. Oceanic crust is formed at sea-floor spreading centers, and continental crust is formed through arc volcanism and accretion of terranes through tectonic processes, though some of these terranes may contain ophiolite sequences, which are pieces of oceanic crust considered to be part of the continent when they exit the standard cycle of formation and spreading centers and subduction beneath continents. Oceanic crust is also denser than continental crust owing to their different compositions. Oceanic crust is denser because it has less silicon and more heavier elements ("mafic") than continental crust ("felsic"). As a result of this density stratification, oceanic crust generally lies below sea level (for example most of the Pacific Plate), while continental crust buoyantly projects above sea level.

Types of Plate Boundaries

Three types of plate boundaries exist, with a fourth, mixed type, characterized by the way the plates move relative to each other. They are associated with different types of surface phenomena. The different types of plate boundaries are:

1. *Transform boundaries (Conservative)* occur where two lithospheric plates slide, or perhaps more accurately, grind past each other along transform faults, where plates are neither created nor destroyed. The relative motion of the two plates is either sinistral (left side toward the observer) or dextral (right side toward the observer). Transform faults occur across a spreading center. Strong earthquakes can occur along a fault. The San Andreas Fault in California is an example of a transform boundary exhibiting dextral motion.

2. *Divergent boundaries (Constructive)* occur where two plates slide apart from each other. At zones of ocean-to-ocean rifting, divergent boundaries form by seafloor spreading, allowing for the formation of new ocean basin. As the continent splits, the ridge forms at the spreading center, the ocean basin expands, and finally, the plate area increases causing many small volcanoes and/or shallow earthquakes. At zones of continent-to-continent rifting, divergent boundaries may cause new ocean basin to form as the continent splits, spreads, the central rift collapses, and ocean fills the basin. Active zones of Mid-ocean ridges (e.g., Mid-Atlantic Ridge and East Pacific Rise), and continent-to-continent rifting (such as Africa's East African Rift and Valley, Red Sea) are examples of divergent boundaries.

3. *Convergent boundaries (Destructive)* (or *active margins*) occur where two plates slide toward each other to form either a subduction zone (one plate moving underneath the other) or a continental collision. At zones of ocean-to-continent subduction (e.g. the Andes mountain range in South America, and the Cascade Mountains in Western United States), the dense oceanic lithosphere plunges beneath the less dense continent. Earthquakes trace the path of the downward-moving plate as it descends into asthenosphere, a trench forms, and as the subducted plate is heated it releases volatiles, mostly water from hydrous minerals, into the surrounding mantle. The addition of water lowers the melting point of the mantle material above the subducting slab, causing it to melt. The magma that results typically leads to volcanism. At zones of ocean-to-ocean subduction (e.g. Aleutian islands, Mariana Islands, and the Japanese island arc), older, cooler, denser crust slips beneath less dense crust. This causes earthquakes and a deep trench to form in an arc shape. The upper mantle of the subducted plate then heats and magma rises to form curving chains of volcanic islands. Deep marine trenches are typically associated with subduction zones, and the basins that develop along the active boundary are often called "foreland basins". Closure of ocean basins can occur at continent-to-continent boundaries (e.g., Himalayas and Alps): collision between masses of granitic continental lithosphere; neither mass is subducted; plate edges are compressed, folded, uplifted.

4. *Plate boundary zones* occur where the effects of the interactions are unclear, and the boundaries, usually occurring along a broad belt, are not well defined and may show various types of movements in different episodes.

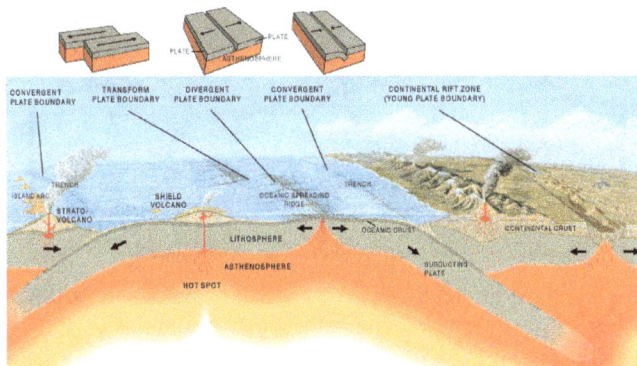

Three types of plate boundary.

Driving Forces of Plate Motion

It is generally accepted that tectonic plates are able to move because of the relative density of oceanic lithosphere and the relative weakness of the asthenosphere. Dissipation of heat from the mantle is acknowledged to be the original source of the energy required to drive plate tectonics through convection or large scale upwelling and doming. The current view, though still a matter of some debate, asserts that as a consequence, a powerful source of plate motion is generated due to the excess density of the oceanic lithosphere sinking in subduction zones. When the new crust forms at mid-ocean ridges, this oceanic lithosphere is initially less dense than the underlying asthenosphere, but it becomes denser with age as it conductively cools and thickens. The greater density of old lithosphere relative to the underlying asthenosphere allows it to sink into the deep mantle at subduction zones, providing most of the driving force for plate movement. The weakness of the asthenosphere allows the tectonic plates to move easily towards a subduction zone. Although subduction is thought to be the strongest force driving plate motions, it cannot be the only force since there are plates such as the North American Plate which are moving, yet are nowhere being subducted. The same is true for the enormous Eurasian Plate. The sources of plate motion are a matter of intensive research and discussion among scientists. One of the main points is that the kinematic pattern of the movement itself should be separated clearly from the possible geodynamic mechanism that is invoked as the driving force of the observed movement, as some patterns may be explained by more than one mechanism. In short, the driving forces advocated at the moment can be divided into three categories based on the relationship to the movement: mantle dynamics related, gravity related (mostly secondary forces).

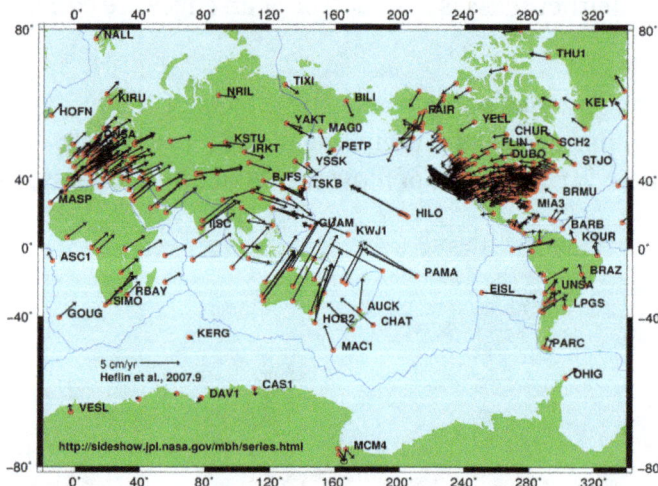

Plate motion based on Global Positioning System (GPS) satellite data from NASA JPL. The vectors show direction and magnitude of motion.

Driving Forces Related to Mantle Dynamics

For much of the last quarter century, the leading theory of the driving force behind tectonic plate motions envisaged large scale convection currents in the upper mantle which are transmitted through the asthenosphere. This theory was launched by Arthur Holmes and some forerunners in the 1930s and was immediately recognized as the solution for the acceptance of the theory as originally discussed in the papers of Alfred Wegener in the early years of the century. However, despite its acceptance, it was long debated in the scientific community because the leading ("fixist") theory still envisaged a static Earth without moving continents up until the major breakthroughs of the early sixties.

Two- and three-dimensional imaging of Earth's interior (seismic tomography) shows a varying lateral density distribution throughout the mantle. Such density variations can be material (from rock chemistry), mineral (from variations in mineral structures), or thermal (through thermal expansion and contraction from heat energy). The manifestation of this varying lateral density is mantle convection from buoyancy forces.

How mantle convection directly and indirectly relates to plate motion is a matter of ongoing study and discussion in geodynamics. Somehow, this energy must be transferred to the lithosphere for tectonic plates to move. There are essentially two types of forces that are thought to influence plate motion: friction and gravity.

- Basal drag (friction): Plate motion driven by friction between the convection currents in the asthenosphere and the more rigid overlying lithosphere.

- Slab suction (gravity): Plate motion driven by local convection currents that exert a downward pull on plates in subduction zones at ocean trenches. Slab suction may occur in a geodynamic setting where basal tractions continue to act on the plate as it dives into the mantle (although perhaps to a greater extent acting on both the under and upper side of the slab).

Lately, the convection theory has been much debated as modern techniques based on 3D seismic tomography still fail to recognize these predicted large scale convection cells. Therefore, alternative views have been proposed:

In the theory of plume tectonics developed during the 1990s, a modified concept of mantle convection currents is used. It asserts that super plumes rise from the deeper mantle and are the drivers or substitutes of the major convection cells. These ideas, which find their roots in the early 1930s with the so-called "fixistic" ideas of the European and Russian Earth Science Schools, find resonance in the modern theories which envisage hot spots/mantle plumes which remain fixed and are overridden by oceanic and continental lithosphere plates over time and leave their traces in the geological record (though these phenomena are not invoked as real driving mechanisms, but rather as modulators). Modern theories that continue building on the older mantle doming concepts and see plate movements as a secondary phenomena are beyond the scope of this page and are discussed elsewhere (for example on the plume tectonics page).

Another theory is that the mantle flows neither in cells nor large plumes but rather as a series of channels just below the Earth's crust, which then provide basal friction to the lithosphere. This theory, called "surge tectonics", became quite popular in geophysics and geodynamics during the 1980s and 1990s.

Driving Forces Related to Gravity

Forces related to gravity are usually invoked as secondary phenomena within the framework of a more general driving mechanism such as the various forms of mantle dynamics described above.

Gravitational sliding away from a spreading ridge: According to many authors, plate motion is driven by the higher elevation of plates at ocean ridges. As oceanic lithosphere is formed at spreading ridges from hot mantle material, it gradually cools and thickens with age (and thus adds distance from the ridge). Cool oceanic lithosphere is significantly denser than the hot mantle material from which it is derived and so with increasing thickness it gradually subsides into the mantle to compensate the greater load. The result is a slight lateral incline with increased distance from the ridge axis.

This force is regarded as a secondary force and is often referred to as "ridge push". This is a misnomer as nothing is "pushing" horizontally and tensional features are dominant along ridges. It is more accurate to refer to this mechanism as gravitational sliding as variable topography across the totality of the plate can vary considerably and the topography of spreading ridges is only the most prominent feature. Other mechanisms generating this gravitational secondary force include flexural bulging of the lithosphere before it dives underneath an adjacent plate which produces a clear topographical feature that can offset, or at least affect, the influence of topographical ocean ridges, and mantle plumes and hot spots, which are postulated to impinge on the underside of tectonic plates.

Slab-pull: Current scientific opinion is that the asthenosphere is insufficiently competent or rigid to directly cause motion by friction along the base of the lithosphere. Slab pull is therefore most widely thought to be the greatest force acting on the plates. In this current understanding, plate motion is mostly driven by the weight of cold, dense plates sinking into the mantle at trenches. Recent models indicate that trench suction plays an important role as well. However, as the North American Plate is nowhere being subducted, yet it is in motion presents a problem. The same holds for the African, Eurasian, and Antarctic plates.

Gravitational sliding away from mantle doming: According to older theories, one of the driving mechanisms of the plates is the existence of large scale asthenosphere/mantle domes which cause the gravitational sliding of lithosphere plates away from them. This gravitational sliding represents a secondary phenomenon of this basically vertically oriented mechanism. This can act on various scales, from the small scale of one island arc up to the larger scale of an entire ocean basin.

Driving Forces Related to Earth Rotation

Alfred Wegener, being a meteorologist, had proposed tidal forces and pole flight force as the main driving mechanisms behind continental drift; however, these forces were considered far too small to cause continental motion as the concept then was of continents plowing through oceanic crust. Therefore, Wegener later changed his position and asserted that convection currents are the main driving force of plate tectonics in the last edition of his book in 1929.

However, in the plate tectonics context (accepted since the seafloor spreading proposals of Heezen, Hess, Dietz, Morley, Vine, and Matthews during the early 1960s), oceanic crust is suggested to be in motion *with* the continents which caused the proposals related to Earth rotation to be reconsidered. In more recent literature, these driving forces are:

1. Tidal drag due to the gravitational force the Moon (and the Sun) exerts on the crust of the Earth

2. Global deformation of the geoid due to small displacements of rotational pole with respect to the Earth's crust;

3. Other smaller deformation effects of the crust due to wobbles and spin movements of the Earth rotation on a smaller time scale.

Forces that are small and generally negligible are:

1. The Coriolis force

2. The centrifugal force, which is treated as a slight modification of gravity

For these mechanisms to be overall valid, systematic relationships should exist all over the globe between the orientation and kinematics of deformation and the geographical latitudinal and longitudinal grid of the Earth itself. Ironically, these systematic relations studies in the second half of the nineteenth century and the first half of the twentieth century underline exactly the opposite: that the plates had not moved in time, that the deformation grid was fixed with respect to the Earth equator and axis, and that gravitational driving forces were generally acting vertically and caused only local horizontal movements (the so-called pre-plate tectonic, "fixist theories"). Later studies (discussed below on this page), therefore, invoked many of the relationships recognized during this pre-plate tectonics period to support their theories.

Of the many forces discussed in this paragraph, tidal force is still highly debated and defended as a possible principle driving force of plate tectonics. The other forces are only used in global geodynamic models not using plate tectonics concepts (therefore beyond the discussions treated in this section) or proposed as minor modulations within the overall plate tectonics model.

In 1973, George W. Moore of the USGS and R. C. Bostrom presented evidence for a general westward drift of the Earth's lithosphere with respect to the mantle. He concluded that tidal forces (the tidal lag or "friction") caused by the Earth's rotation and the forces acting upon it by the Moon are a driving force for plate tectonics. As the Earth spins eastward beneath the moon, the moon's gravity ever so slightly pulls the Earth's surface layer back westward, just as proposed by Alfred Wegener. In a more recent 2006 study, scientists reviewed and advocated these earlier proposed ideas. It has also been suggested recently in Lovett (2006) that this observation may also explain why Venus and Mars have no plate tectonics, as Venus has no moon and Mars' moons are too small to have significant tidal effects on the planet. In a recent paper, it was suggested that, on the other hand, it can easily be observed that many plates are moving north and eastward, and that the dominantly westward motion of the Pacific ocean basins derives simply from the eastward bias of the Pacific spreading center (which is not a predicted manifestation of such lunar forces). In the same paper the authors admit, however, that relative to the lower mantle, there is a slight westward component in the motions of all the plates. They demonstrated though that the westward drift, seen only for the past 30 Ma, is attributed to the increased dominance of the steadily growing and accelerating Pacific plate. The debate is still open.

Relative Significance of Each Driving Force Mechanism

The vector of a plate's motion is a function of all the forces acting on the plate; however, therein lies the problem regarding the degree to which each process contributes to the overall motion of each tectonic plate.

The diversity of geodynamic settings and the properties of each plate result from the impact of the various processes actively driving each individual plate. One method of dealing with this problem is to consider the relative rate at which each plate is moving as well as the evidence related to the significance of each process to the overall driving force on the plate.

One of the most significant correlations discovered to date is that lithospheric plates attached to downgoing (subducting) plates move much faster than plates not attached to subducting plates. The Pacific plate, for instance, is essentially surrounded by zones of subduction (the so-called Ring of Fire) and moves much faster than the plates of the Atlantic basin, which are attached (perhaps one could say 'welded') to adjacent continents instead of subducting plates. It is thus thought that forces associated with the downgoing plate (slab pull and slab suction) are the driving forces which determine the motion of plates, except for those plates which are not being subducted. The driving forces of plate motion continue to be active subjects of on-going research within geophysics and tectonophysics.

Development of the Theory

Summary

In line with other previous and contemporaneous proposals, in 1912 the meteorologist Alfred Wegener amply described what he called continental drift, expanded in his 1915 book *The Origin of Continents and Oceans* and the scientific debate started that would end up fifty years later in the theory of plate tectonics. Starting from the idea (also expressed by his forerunners) that the present continents once formed a single land mass (which was called Pangea later on) that drifted apart, thus releasing the continents from the Earth's mantle and likening them to "icebergs" of low density granite floating on a sea of denser basalt. Supporting evidence for the idea came from the dove-tailing outlines of South America's east coast and Africa's west coast, and from the matching of the rock formations along these edges. Confirmation of their previous contiguous nature also came from the fossil plants *Glossopteris* and *Gangamopteris*, and the therapsid or mammal-like reptile *Lystrosaurus*, all widely distributed over South America, Africa, Antarctica, India and Australia. The evidence for such an erstwhile joining of these continents was patent to field geologists working in the southern hemisphere. The South African Alex du Toit put together a mass of such information in his 1937 publication *Our Wandering Continents*, and went further than Wegener in recognising the strong links between the Gondwana fragments.

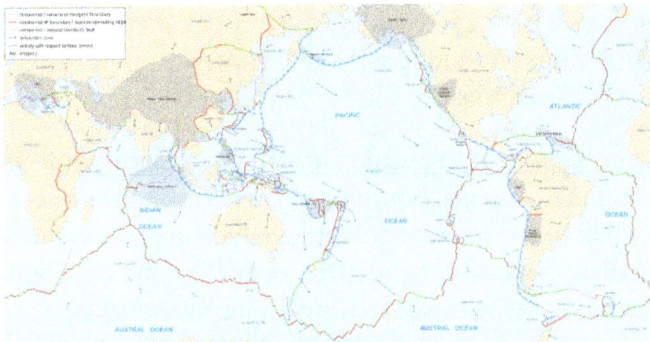

Detailed map showing the tectonic plates with their movement vectors.

But without detailed evidence and a force sufficient to drive the movement, the theory was not generally accepted: the Earth might have a solid crust and mantle and a liquid core, but there seemed to be no way that portions of the crust could move around. Distinguished scientists, such as Harold Jeffreys and Charles Schuchert, were outspoken critics of continental drift.

Despite much opposition, the view of continental drift gained support and a lively debate started between "drifters" or "mobilists" (proponents of the theory) and "fixists" (opponents). During the 1920s, 1930s and 1940s, the former reached important milestones proposing that convection currents might have driven the plate movements, and that spreading may have occurred below the sea within the oceanic crust. Concepts close to the elements now incorporated in plate tectonics were proposed by geophysicists and geologists (both fixists and mobilists) like Vening-Meinesz, Holmes, and Umbgrove.

One of the first pieces of geophysical evidence that was used to support the movement of lithospheric plates came from paleomagnetism. This is based on the fact that rocks of different ages show a variable magnetic field direction, evidenced by studies since the mid–nineteenth century. The magnetic north and south poles reverse through time, and, especially important in paleotectonic studies, the relative position of the magnetic north pole varies through time. Initially, during the first half of the twentieth century, the latter phenomenon was explained by introducing what was called "polar wander", i.e., it was assumed that the north pole location had been shifting through time. An alternative explanation, though, was that the continents had moved (shifted and rotated) relative to the north pole, and each continent, in fact, shows its own "polar wander path". During the late 1950s it was successfully shown on two occasions that these data could show the validity of continental drift: by Keith Runcorn in a paper in 1956, and by Warren Carey in a symposium held in March 1956.

The second piece of evidence in support of continental drift came during the late 1950s and early 60s from data on the bathymetry of the deep ocean floors and the nature of the oceanic crust such as magnetic properties and, more generally, with the development of marine geology which gave evidence for the association of seafloor spreading along the mid-oceanic ridges and magnetic field reversals, published between 1959 and 1963 by Heezen, Dietz, Hess, Mason, Vine & Matthews, and Morley.

Simultaneous advances in early seismic imaging techniques in and around Wadati-Benioff zones along the trenches bounding many continental margins, together with many other geophysical (e.g. gravimetric) and geological observations, showed how the oceanic crust could disappear into the mantle, providing the mechanism to balance the extension of the ocean basins with shortening along its margins.

All this evidence, both from the ocean floor and from the continental margins, made it clear around 1965 that continental drift was feasible and the theory of plate tectonics, which was defined in a series of papers between 1965 and 1967, was born, with all its extraordinary explanatory and predictive power. The theory revolutionized the Earth sciences, explaining a diverse range of geological phenomena and their implications in other studies such as paleogeography and palcobiology.

Continental Drift

In the late 19th and early 20th centuries, geologists assumed that the Earth's major features were fixed, and that most geologic features such as basin development and mountain ranges could be explained by vertical crustal movement, described in what is called the geosynclinal theory. Generally, this was placed in the context of a contracting planet Earth due to heat loss in the course of a relatively short geological time.

It was observed as early as 1596 that the opposite coasts of the Atlantic Ocean—or, more precisely, the edges of the continental shelves—have similar shapes and seem to have once fitted together.

Alfred Wegener in Greenland in the winter of 1912-13.

Since that time many theories were proposed to explain this apparent complementarity, but the assumption of a solid Earth made these various proposals difficult to accept.

The discovery of radioactivity and its associated heating properties in 1895 prompted a re-examination of the apparent age of the Earth. This had previously been estimated by its cooling rate and assumption the Earth's surface radiated like a black body. Those calculations had implied that, even if it started at red heat, the Earth would have dropped to its present temperature in a few tens of millions of years. Armed with the knowledge of a new heat source, scientists realized that the Earth would be much older, and that its core was still sufficiently hot to be liquid.

By 1915, after having published a first article in 1912, Alfred Wegener was making serious arguments for the idea of continental drift in the first edition of *The Origin of Continents and Oceans*. In that book (re-issued in four successive editions up to the final one in 1936), he noted how the east coast of South America and the west coast of Africa looked as if they were once attached. Wegener was not the first to note this (Abraham Ortelius, Antonio Snider-Pellegrini, Eduard Suess, Roberto Mantovani and Frank Bursley Taylor preceded him just to mention a few), but he was the first to marshal significant fossil and paleo-topographical and climatological evidence to support this simple observation (and was supported in this by researchers such as Alex du Toit). Furthermore, when the rock strata of the margins of separate continents are very similar it suggests that these rocks were formed in the same way, implying that they were joined initially. For instance, parts of Scotland and Ireland contain rocks very similar to those found in Newfoundland and New Brunswick. Furthermore, the Caledonian Mountains of Europe and parts of the Appalachian Mountains of North America are very similar in structure and lithology.

However, his ideas were not taken seriously by many geologists, who pointed out that there was no apparent mechanism for continental drift. Specifically, they did not see how continental rock could plow through the much denser rock that makes up oceanic crust. Wegener could not explain the force that drove continental drift, and his vindication did not come until after his death in 1930.

Floating Continents, Paleomagnetism, and Seismicity Zones

As it was observed early that although granite existed on continents, seafloor seemed to be composed of denser basalt, the prevailing concept during the first half of the twentieth century was that there were two types of crust, named "sial" (continental type crust) and "sima" (oceanic type crust). Furthermore, it was supposed that a static shell of strata was present under the continents. It therefore looked apparent that a layer of basalt (sial) underlies the continental rocks.

Preliminary Determination of Epicenters
358,214 Events, 1963 - 1998

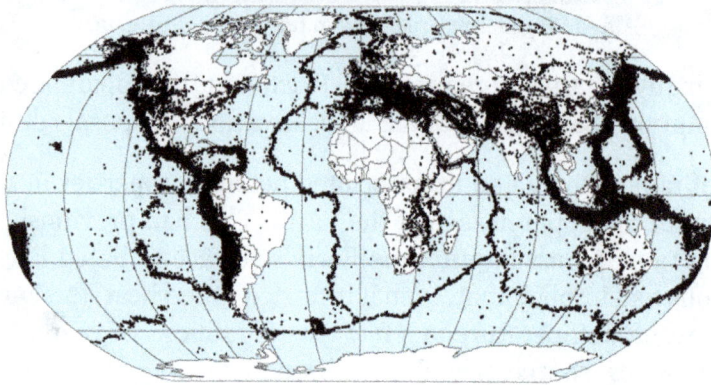

Global earthquake epicenters, 1963–1998

However, based on abnormalities in plumb line deflection by the Andes in Peru, Pierre Bouguer had deduced that less-dense mountains must have a downward projection into the denser layer underneath. The concept that mountains had "roots" was confirmed by George B. Airy a hundred years later, during study of Himalayan gravitation, and seismic studies detected corresponding density variations. Therefore, by the mid-1950s, the question remained unresolved as to whether mountain roots were clenched in surrounding basalt or were floating on it like an iceberg.

During the 20th century, improvements in and greater use of seismic instruments such as seismographs enabled scientists to learn that earthquakes tend to be concentrated in specific areas, most notably along the oceanic trenches and spreading ridges. By the late 1920s, seismologists were beginning to identify several prominent earthquake zones parallel to the trenches that typically were inclined 40–60° from the horizontal and extended several hundred kilometers into the Earth. These zones later became known as Wadati-Benioff zones, or simply Benioff zones, in honor of the seismologists who first recognized them, Kiyoo Wadati of Japan and Hugo Benioff of the United States. The study of global seismicity greatly advanced in the 1960s with the establishment of the Worldwide Standardized Seismograph Network (WWSSN) to monitor the compliance of the 1963 treaty banning above-ground testing of nuclear weapons. The much improved data from the WWSSN instruments allowed seismologists to map precisely the zones of earthquake concentration worldwide.

Meanwhile, debates developed around the phenomena of polar wander. Since the early debates of continental drift, scientists had discussed and used evidence that polar drift had occurred because continents seemed to have moved through different climatic zones during the past. Furthermore, paleomagnetic data had shown that the magnetic pole had also shifted during time. Reasoning in an opposite way, the continents might have shifted and rotated, while the pole remained relatively fixed. The first time the evidence of magnetic polar wander was used to support the movements of continents was in a paper by Keith Runcorn in 1956, and successive papers by him and his students Ted Irving (who was actually the first to be convinced of the fact that paleomagnetism supported continental drift) and Ken Creer.

This was immediately followed by a symposium in Tasmania in March 1956. In this symposium, the evidence was used in the theory of an expansion of the global crust. In this hypothesis the shifting of the continents can be simply explained by a large increase in size of the Earth since its formation. However, this was unsatisfactory because its supporters could offer no convincing mechanism to produce a significant expansion of the Earth. Certainly there is no evidence that the moon has expanded in the past 3 billion years; other work would soon show that the evidence was equally in support of continental drift on a globe with a stable radius.

During the thirties up to the late fifties, works by Vening-Meinesz, Holmes, Umbgrove, and numerous others outlined concepts that were close or nearly identical to modern plate tectonics theory. In particular, the English geologist Arthur Holmes proposed in 1920 that plate junctions might lie beneath the sea, and in 1928 that convection currents within the mantle might be the driving force. Often, these contributions are forgotten because:

- At the time, continental drift was not accepted.

- Some of these ideas were discussed in the context of abandoned fixistic ideas of a deforming globe without continental drift or an expanding Earth.

- They were published during an episode of extreme political and economic instability that hampered scientific communication.

- Many were published by European scientists and at first not mentioned or given little credit in the papers on sea floor spreading published by the American researchers in the 1960s.

Mid-oceanic Ridge Spreading and Convection

In 1947, a team of scientists led by Maurice Ewing utilizing the Woods Hole Oceanographic Institution's research vessel *Atlantis* and an array of instruments, confirmed the existence of a rise in the central Atlantic Ocean, and found that the floor of the seabed beneath the layer of sediments consisted of basalt, not the granite which is the main constituent of continents. They also found that the oceanic crust was much thinner than continental crust. All these new findings raised important and intriguing questions.

The new data that had been collected on the ocean basins also showed particular characteristics regarding the bathymetry. One of the major outcomes of these datasets was that all along the globe, a system of mid-oceanic ridges was detected. An important conclusion was that along this system, new ocean floor was being created, which led to the concept of the "Great Global Rift". This was described in the crucial paper of Bruce Heezen (1960), which would trigger a real revolution in thinking. A profound consequence of seafloor spreading is that new crust was, and still is, being continually created along the oceanic ridges. Therefore, Heezen advocated the so-called "expanding Earth" hypothesis of S. Warren Carey. So, still the question remained: how can new crust be continuously added along the oceanic ridges without increasing the size of the Earth? In reality, this question had been solved already by numerous scientists during the forties and the fifties, like Arthur Holmes, Vening-Meinesz, Coates and many others: The crust in excess disappeared along what were called the oceanic trenches, where so-called "subduction" occurred. Therefore, when various scientists during the early sixties started to reason on the data at their disposal regarding the ocean floor, the pieces of the theory quickly fell into place.

The question particularly intrigued Harry Hammond Hess, a Princeton University geologist and a Naval Reserve Rear Admiral, and Robert S. Dietz, a scientist with the U.S. Coast and Geodetic Survey who first coined the term *seafloor spreading*. Dietz and Hess (the former published the same idea one year earlier in *Nature*, but priority belongs to Hess who had already distributed an unpublished manuscript of his 1962 article by 1960) were among the small handful who really understood the broad implications of sea floor spreading and how it would eventually agree with the, at that time, unconventional and unaccepted ideas of continental drift and the elegant and mobilistic models proposed by previous workers like Holmes.

In the same year, Robert R. Coats of the U.S. Geological Survey described the main features of island arc subduction in the Aleutian Islands. His paper, though little noted (and even ridiculed) at the time, has since been called "seminal" and "prescient". In reality, it actually shows that the work by the European scientists on island arcs and mountain belts performed and published during the 1930s up until the 1950s was applied and appreciated also in the United States.

If the Earth's crust was expanding along the oceanic ridges, Hess and Dietz reasoned like Holmes and others before them, it must be shrinking elsewhere. Hess followed Heezen, suggesting that new oceanic crust continuously spreads away from the ridges in a conveyor belt–like motion. And, using the mobilistic concepts developed before, he correctly concluded that many millions of years later, the oceanic crust eventually descends along the continental margins where oceanic trenches – very deep, narrow canyons – are formed, e.g. along the rim of the Pacific Ocean basin. The important step Hess made was that convection currents would be the driving force in this process, arriving at the same conclusions as Holmes had decades before with the only difference that the thinning of the ocean crust was performed using Heezen's mechanism of spreading along the ridges. Hess therefore concluded that the Atlantic Ocean was expanding while the Pacific Ocean was shrinking. As old oceanic crust is "consumed" in the trenches (like Holmes and others, he thought this was done by thickening of the continental lithosphere, not, as now understood, by underthrusting at a larger scale of the oceanic crust itself into the mantle), new magma rises and erupts along the spreading ridges to form new crust. In effect, the ocean basins are perpetually being "recycled," with the creation of new crust and the destruction of old oceanic lithosphere occurring simultaneously. Thus, the new mobilistic concepts neatly explained why the Earth does not get bigger with sea floor spreading, why there is so little sediment accumulation on the ocean floor, and why oceanic rocks are much younger than continental rocks.

Magnetic Striping

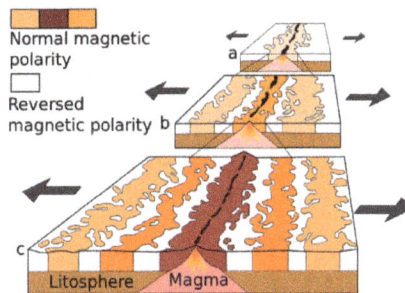

Seafloor magnetic striping.

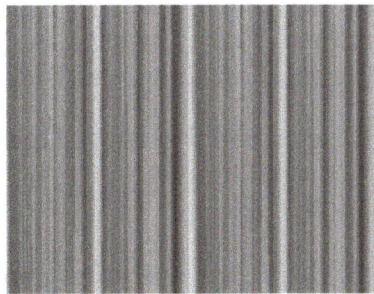

A demonstration of magnetic striping. (The darker the color is, the closer it is to normal polarity)

Beginning in the 1950s, scientists like Victor Vacquier, using magnetic instruments (magnetometers) adapted from airborne devices developed during World War II to detect submarines, began recognizing odd magnetic variations across the ocean floor. This finding, though unexpected, was not entirely surprising because it was known that basalt—the iron-rich, volcanic rock making up the ocean floor—contains a strongly magnetic mineral (magnetite) and can locally distort compass readings. This distortion was recognized by Icelandic mariners as early as the late 18th century. More important, because the presence of magnetite gives the basalt measurable magnetic properties, these newly discovered magnetic variations provided another means to study the deep ocean floor. When newly formed rock cools, such magnetic materials recorded the Earth's magnetic field at the time.

As more and more of the seafloor was mapped during the 1950s, the magnetic variations turned out not to be random or isolated occurrences, but instead revealed recognizable patterns. When these magnetic patterns were mapped over a wide region, the ocean floor showed a zebra-like pattern: one stripe with normal polarity and the adjoining stripe with reversed polarity. The overall pattern, defined by these alternating bands of normally and reversely polarized rock, became known as magnetic striping, and was published by Ron G. Mason and co-workers in 1961, who did not find, though, an explanation for these data in terms of sea floor spreading, like Vine, Matthews and Morley a few years later.

The discovery of magnetic striping called for an explanation. In the early 1960s scientists such as Heezen, Hess and Dietz had begun to theorise that mid-ocean ridges mark structurally weak zones where the ocean floor was being ripped in two lengthwise along the ridge crest. New magma from deep within the Earth rises easily through these weak zones and eventually erupts along the crest of the ridges to create new oceanic crust. This process, at first denominated the "conveyer belt hypothesis" and later called seafloor spreading, operating over many millions of years continues to form new ocean floor all across the 50,000 km-long system of mid-ocean ridges.

Only four years after the maps with the "zebra pattern" of magnetic stripes were published, the link between sea floor spreading and these patterns was correctly placed, independently by Lawrence Morley, and by Fred Vine and Drummond Matthews, in 1963, now called the Vine-Matthews-Morley hypothesis. This hypothesis linked these patterns to geomagnetic reversals and was supported by several lines of evidence:

1. the stripes are symmetrical around the crests of the mid-ocean ridges; at or near the crest of the ridge, the rocks are very young, and they become progressively older away from the ridge crest;

2. the youngest rocks at the ridge crest always have present-day (normal) polarity;

3. stripes of rock parallel to the ridge crest alternate in magnetic polarity (normal-reversed-normal, etc.), suggesting that they were formed during different epochs documenting the (already known from independent studies) normal and reversal episodes of the Earth's magnetic field.

By explaining both the zebra-like magnetic striping and the construction of the mid-ocean ridge system, the seafloor spreading hypothesis (SFS) quickly gained converts and represented another major advance in the development of the plate-tectonics theory.

Furthermore, the oceanic crust now came to be appreciated as a natural "tape recording" of the history of the geomagnetic field reversals (GMFR) of the Earth's magnetic field. Today, extensive studies are dedicated to the calibration of the normal-reversal patterns in the oceanic crust on one hand and known timescales derived from the dating of basalt layers in sedimentary sequences (magnetostratigraphy) on the other, to arrive at estimates of past spreading rates and plate reconstructions.

Definition and Refining of the Theory

After all these considerations, Plate Tectonics (or, as it was initially called "New Global Tectonics") became quickly accepted in the scientific world, and numerous papers followed that defined the concepts:

- In 1965, Tuzo Wilson who had been a promotor of the sea floor spreading hypothesis and continental drift from the very beginning added the concept of transform faults to the model, completing the classes of fault types necessary to make the mobility of the plates on the globe work out.

- A symposium on continental drift was held at the Royal Society of London in 1965 which must be regarded as the official start of the acceptance of plate tectonics by the scientific community, and which abstracts are issued as Blacket, Bullard & Runcorn (1965). In this symposium, Edward Bullard and co-workers showed with a computer calculation how the continents along both sides of the Atlantic would best fit to close the ocean, which became known as the famous "Bullard's Fit".

- In 1966 Wilson published the paper that referred to previous plate tectonic reconstructions, introducing the concept of what is now known as the "Wilson Cycle".

- In 1967, at the American Geophysical Union's meeting, W. Jason Morgan proposed that the Earth's surface consists of 12 rigid plates that move relative to each other.

- Two months later, Xavier Le Pichon published a complete model based on 6 major plates with their relative motions, which marked the final acceptance by the scientific community of plate tectonics.

- In the same year, McKenzie and Parker independently presented a model similar to Morgan's using translations and rotations on a sphere to define the plate motions.

Implications for Biogeography

Continental drift theory helps biogeographers to explain the disjunct biogeographic distribution of present-day life found on different continents but having similar ancestors. In particular, it explains the Gondwanan distribution of ratites and the Antarctic flora.

Plate Reconstruction

Reconstruction is used to establish past (and future) plate configurations, helping determine the shape and make-up of ancient supercontinents and providing a basis for paleogeography.

Defining Plate Boundaries

Current plate boundaries are defined by their seismicity. Past plate boundaries within existing plates are identified from a variety of evidence, such as the presence of ophiolites that are indicative of vanished oceans.

Past Plate Motions

Tectonic motion first began around three billion years ago.

Various types of quantitative and semi-quantitative information are available to constrain past plate motions. The geometric fit between continents, such as between west Africa and South America is still an important part of plate reconstruction. Magnetic stripe patterns provide a reliable guide to relative plate motions going back into the Jurassic period. The tracks of hotspots give absolute reconstructions, but these are only available back to the Cretaceous. Older reconstructions rely mainly on paleomagnetic pole data, although these only constrain the latitude and rotation, but not the longitude. Combining poles of different ages in a particular plate to produce apparent polar wander paths provides a method for comparing the motions of different plates through time. Additional evidence comes from the distribution of certain sedimentary rock types, faunal provinces shown by particular fossil groups, and the position of orogenic belts.

Formation and Break-up of Continents

The movement of plates has caused the formation and break-up of continents over time, including occasional formation of a supercontinent that contains most or all of the continents. The supercontinent Columbia or Nuna formed during a period of 2,000 to 1,800 million years ago and broke up about 1,500 to 1,300 million years ago. The supercontinent Rodinia is thought to have formed about 1 billion years ago and to have embodied most or all of Earth's continents, and broken up into eight continents around 600 million years ago. The eight continents later re-assembled into another supercontinent called Pangaea; Pangaea broke up into Laurasia (which became North America and Eurasia) and Gondwana (which became the remaining continents).

The Himalayas, the world's tallest mountain range, are assumed to have been formed by the collision of two major plates. Before uplift, they were covered by the Tethys Ocean.

Current Plates

There are dozens of smaller plates, the seven largest of which are the Arabian, Caribbean, Juan de Fuca, Cocos, Nazca, Philippine Sea and Scotia.

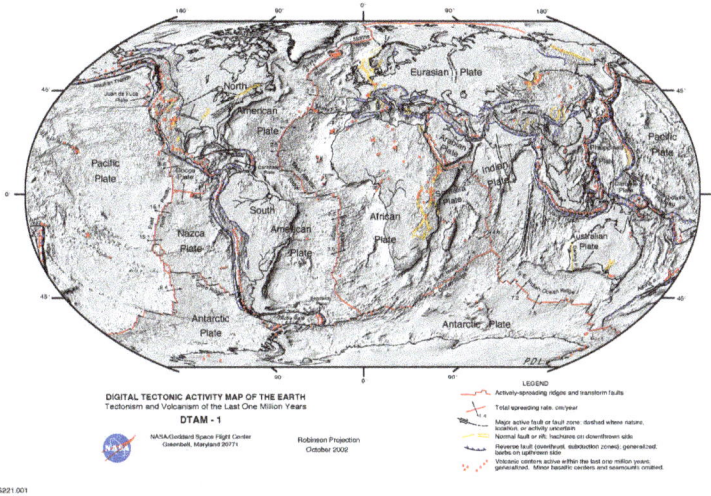

DIGITAL TECTONIC ACTIVITY MAP OF THE EARTH
Tectonism and Volcanism of the Last One Million Years
DTAM - 1

NASA/Goddard Space Flight Center
Greenbelt, Maryland 20771

Robinson Projection
October 2002

Depending on how they are defined, there are usually seven or eight "major" plates: African, Antarctic, Eurasian, North American, South American, Pacific, and Indo-Australian. The latter is sometimes subdivided into the Indian and Australian plates.

The current motion of the tectonic plates is today determined by remote sensing satellite data sets, calibrated with ground station measurements.

Other Celestial Bodies (Planets, Moons)

The appearance of plate tectonics on terrestrial planets is related to planetary mass, with more massive planets than Earth expected to exhibit plate tectonics. Earth may be a borderline case, owing its tectonic activity to abundant water (silica and water form a deep eutectic.)

Venus

Venus shows no evidence of active plate tectonics. There is debatable evidence of active tectonics in the planet's distant past; however, events taking place since then (such as the plausible and generally accepted hypothesis that the Venusian lithosphere has thickened greatly over the course of several hundred million years) has made constraining the course of its geologic record difficult. However, the numerous well-preserved impact craters have been utilized as a dating method to approximately date the Venusian surface (since there are thus far no known samples of Venusian rock to be dated by more reliable methods). Dates derived are dominantly in the range 500 to 750 million years ago, although ages of up to 1,200 million years ago have been calculated. This research has led to the fairly well accepted hypothesis that Venus has undergone an essentially complete volcanic resurfacing at least once in its distant past, with the last event taking place approximately within the range of estimated surface ages. While the mechanism of such an impressive thermal event remains a debated issue in Venusian geosciences, some scientists are advocates of processes involving plate motion to some extent.

One explanation for Venus's lack of plate tectonics is that on Venus temperatures are too high for significant water to be present. The Earth's crust is soaked with water, and water plays an important role in the development of shear zones. Plate tectonics requires weak surfaces in the crust along which crustal slices can move, and it may well be that such weakening never took place on Venus because of the absence of water. However, some researchers remain convinced that plate tectonics is or was once active on this planet.

Mars

Mars is considerably smaller than Earth and Venus, and there is evidence for ice on its surface and in its crust.

In the 1990s, it was proposed that Martian Crustal Dichotomy was created by plate tectonic processes. Scientists today disagree, and think that it was created either by up-welling within the Martian mantle that thickened the crust of the Southern Highlands and formed Tharsis or by a giant impact that excavated the Northern Lowlands.

Valles Marineris may be a tectonic boundary.

Observations made of the magnetic field of Mars by the *Mars Global Surveyor* space-craft in 1999 showed patterns of magnetic striping discovered on this planet. Some scientists interpreted these as requiring plate tectonic processes, such as seafloor spreading. However, their data fail a "magnetic reversal test", which is used to see if they were formed by flipping polarities of a global magnetic field.

Icy Satellites

Some of the satellites of Jupiter have features that may be related to plate-tectonic style deformation, although the materials and specific mechanisms may be different from plate-tectonic activity on Earth. On 8 September 2014, NASA reported finding evidence of plate tectonics on Europa, a satellite of Jupiter—the first sign of such geological activity on another world other than Earth.

Titan, the largest moon of Saturn, was reported to show tectonic activity in images taken by the *Huygens* probe, which landed on Titan on January 14, 2005.

Exoplanets

On Earth-sized planets, plate tectonics is more likely if there are oceans of water; however, in 2007, two independent teams of researchers came to opposing conclusions about the likelihood of plate tectonics on larger super-earths with one team saying that plate tectonics would be episodic or stagnant and the other team saying that plate tectonics is very likely on super-earths even if the planet is dry.

Geophysics

Geophysics /dʒiːoʊfɪzɪks/ is a subject of natural science concerned with the physical processes and physical properties of the Earth and its surrounding space environment, and the use of quantitative methods for their analysis. The term *geophysics* sometimes refers only to the geological applications: Earth's shape; its gravitational and magnetic fields; its internal structure and composition; its dynamics and their surface expression in plate tectonics, the generation of magmas, volcanism and rock formation. However, modern geophysics organizations use a broader definition that includes the water cycle including snow and ice; fluid dynamics of the oceans and the atmosphere; electricity and magnetism in the ionosphere and magnetosphere and solar-terrestrial relations; and analogous problems associated with the Moon and other planets.

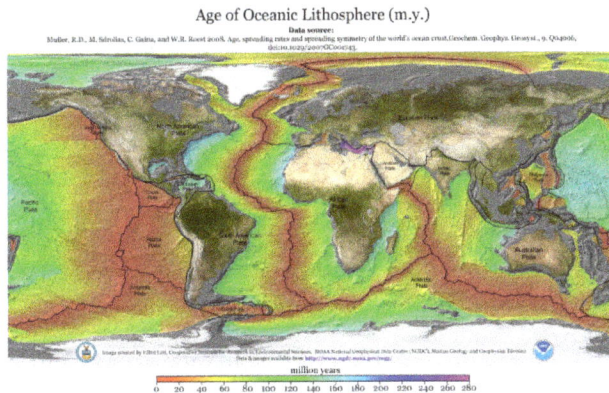

Age of the sea floor. Much of the dating information comes from magnetic anomalies.

Although geophysics was only recognized as a separate discipline in the 19th century, its origins date back to ancient times. The first magnetic compasses were made from lodestones, while more modern magnetic compasses played an important role in the history of navigation. The first seismic instrument was built in 132 BC. Isaac Newton applied his theory of mechanics to the tides and the precession of the equinox; and instruments were developed to measure the Earth's shape, density and gravity field, as well as the components of the water cycle. In the 20th century, geophysical methods were developed for remote exploration of the solid Earth and the ocean, and geophysics played an essential role in the development of the theory of plate tectonics.

Geophysics is applied to societal needs, such as mineral resources, mitigation of natural hazards and environmental protection. Geophysical survey data are used to analyze potential petroleum reservoirs and mineral deposits, locate groundwater, find archaeological relics, determine the thickness of glaciers and soils, and assess sites for environmental remediation.

Physical Phenomena

Geophysics is a highly interdisciplinary subject, and geophysicists contribute to every area of the Earth sciences. To provide a clearer idea of what constitutes geophysics, this section describes phenomena that are studied in physics and how they relate to the Earth and its surroundings.

Gravity

The gravitational pull of the Moon and Sun give rise to two high tides and two low tides every lunar day, or every 24 hours and 50 minutes. Therefore, there is a gap of 12 hours and 25 minutes between every high tide and between every low tide.

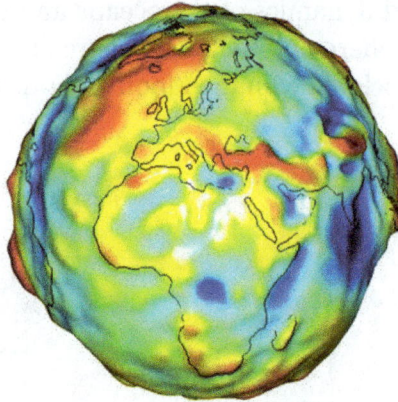

A map of deviations in gravity from a perfectly smooth, idealized Earth.

Gravitational forces make rocks press down on deeper rocks, increasing their density as the depth increases. Measurements of gravitational acceleration and gravitational potential at the Earth's surface and above it can be used to look for mineral deposits. The surface gravitational field provides information on the dynamics of tectonic plates. The geopotential surface called the geoid is one definition of the shape of the Earth. The geoid would be the global mean sea level if the oceans were in equilibrium and could be extended through the continents (such as with very narrow canals).

Heat Flow

The Earth is cooling, and the resulting heat flow generates the Earth's magnetic field through the geodynamo and plate tectonics through mantle convection. The main sources of heat are the primordial heat and radioactivity, although there are also contributions from phase transitions. Heat is mostly carried to the surface by thermal convection, although there are two thermal boundary layers – the core-mantle boundary and the lithosphere – in which heat is transported by conduction. Some heat is carried up from the bottom of the mantle by mantle plumes. The heat flow at the Earth's surface is about 4.2×10^{13} W, and it is a potential source of geothermal energy.

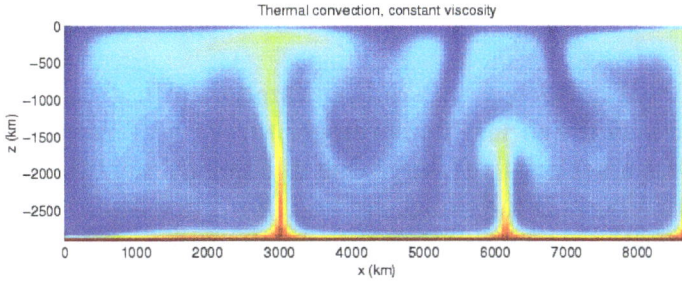

A model of thermal convection in the Earth's mantle. The thin red columns are mantle plumes.

Vibrations

Seismic waves are vibrations that travel through the Earth's interior or along its surface. The entire Earth can also oscillate in forms that are called normal modes or free oscillations of the Earth. Ground motions from waves or normal modes are measured using seismographs. If the waves come from a localized source such as an earthquake or explosion, measurements at more than one location can be used to locate the source. The locations of earthquakes provide information on plate tectonics and mantle convection.

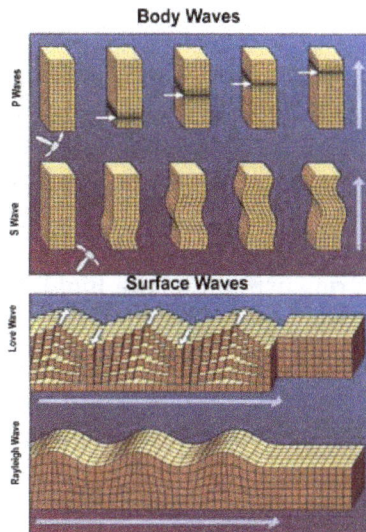

Illustration of the deformations of a block by body waves and surface waves.

Measurements of seismic waves are a source of information on the region that the waves travel through. If the density or composition of the rock changes suddenly, some waves are reflected. Reflections can provide information on near-surface structure. Changes in the travel direction, called refraction, can be used to infer the deep structure of the Earth.

Earthquakes pose a risk to humans. Understanding their mechanisms, which depend on the type of earthquake (e.g., intraplate or deep focus), can lead to better estimates of earthquake risk and improvements in earthquake engineering.

Electricity

Although we mainly notice electricity during thunderstorms, there is always a downward electric field near the surface that averages 120 V m^{-1}. Relative to the solid Earth, the atmosphere has a net positive charge due to bombardment by cosmic rays. A current of about 1800 A flows in the global circuit. It flows downward from the ionosphere over most of the Earth and back upwards through thunderstorms. The flow is manifested by lightning below the clouds and sprites above.

A variety of electric methods are used in geophysical survey. Some measure spontaneous potential, a potential that arises in the ground because of man-made or natural disturbances. Telluric currents flow in Earth and the oceans. They have two causes: electromagnetic induction by the time-varying, external-origin geomagnetic field and motion of conducting bodies (such as seawater) across the Earth's permanent magnetic field. The distribution of telluric current density can be used to detect variations in electrical resistivity of underground structures. Geophysicists can also provide the electric current themselves.

Electromagnetic Waves

Electromagnetic waves occur in the ionosphere and magnetosphere as well as the Earth's outer core. Dawn chorus is believed to be caused by high-energy electrons that get caught in the Van Allen radiation belt. Whistlers are produced by lightning strikes. Hiss may be generated by both. Electromagnetic waves may also be generated by earth-quakes.

In the Earth's outer core, electric currents in the highly conductive liquid iron create magnetic fields by electromagnetic induction. Alfvén waves are magnetohydrodynamic waves in the magnetosphere or the Earth's core. In the core, they probably have little observable effect on the geomagnetic field, but slower waves such as magnetic Rossby waves may be one source of geomagnetic secular variation.

Electromagnetic methods that are used for geophysical survey include transient electromagnetics and magnetotellurics.

Magnetism

The Earth's magnetic field protects the Earth from the deadly solar wind and has long been used for navigation. It originates in the fluid motions of the Earth's outer core. The magnetic field in the upper atmosphere gives rise to the auroras.

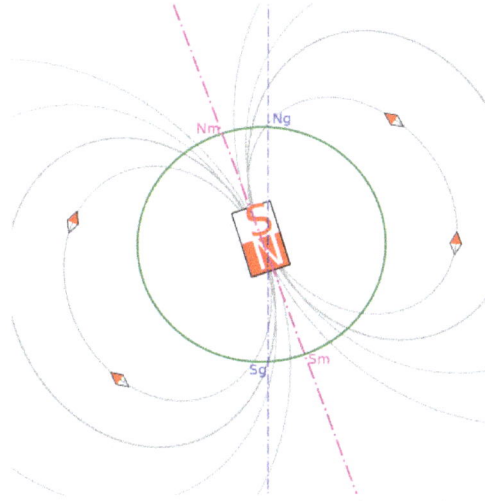

Earth's dipole axis (pink line) is tilted away from the rotational axis (blue line).

The Earth's field is roughly like a tilted dipole, but it changes over time (a phenomenon called geomagnetic secular variation). Mostly the geomagnetic pole stays near the geographic pole, but at random intervals averaging 440,000 to a million years or so, the polarity of the Earth's field reverses. These geomagnetic reversals, analyzed within a Geomagnetic Polarity Time Scale, contain 184 polarity intervals in the last 83 million years, with change in frequency over time, with the most recent brief complete reversal of the Laschamp event occurring 41,000 years ago during the last glacial period. Geologists observed geomagnetic reversal recorded in volcanic rocks, through magnetostratigraphy correlation and their signature can be seen as parallel linear magnetic anomaly stripes on the seafloor. These stripes provide quantitative information on sea-floor spreading, a part of plate tectonics. They are the basis of magnetostratigraphy, which correlates magnetic reversals with other stratigraphies to construct geologic time scales. In addition, the magnetization in rocks can be used to measure the motion of continents.

Radioactivity

Radioactive decay accounts for about 80% of the Earth's internal heat, powering the geodynamo and plate tectonics. The main heat-producing isotopes are potassium-40, uranium-238, uranium-235, and thorium-232. Radioactive elements are used for radiometric dating, the primary method for establishing an absolute time scale in geochronology. Unstable isotopes decay at predictable rates, and the decay rates of different isotopes cover several orders of magnitude, so radioactive decay can be used to accurately date both recent events and events in past geologic eras. Radiometric mapping using ground and airborne gamma spectrometry can be used to map the concentration and distribution of radioisotopes near the Earth's surface, which is useful for mapping lithology and alteration.

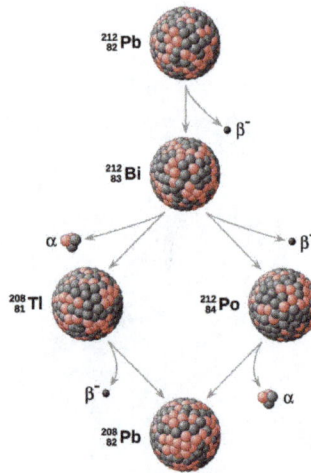

Example of a radioactive decay chain.

Fluid Dynamics

Fluid motions occur in the magnetosphere, atmosphere, ocean, mantle and core. Even the mantle, though it has an enormous viscosity, flows like a fluid over long time intervals. This flow is reflected in phenomena such as isostasy, post-glacial rebound and mantle plumes. The mantle flow drives plate tectonics and the flow in the Earth's core drives the geodynamo.

Geophysical fluid dynamics is a primary tool in physical oceanography and meteorology. The rotation of the Earth has profound effects on the Earth's fluid dynamics, often due to the Coriolis effect. In the atmosphere it gives rise to large-scale patterns like Rossby waves and determines the basic circulation patterns of storms. In the ocean they drive large-scale circulation patterns as well as Kelvin waves and Ekman spirals at the ocean surface. In the Earth's core, the circulation of the molten iron is structured by Taylor columns.

Waves and other phenomena in the magnetosphere can be modeled using magnetohydrodynamics.

Mineral Physics

The physical properties of minerals must be understood to infer the composition of the Earth's interior from seismology, the geothermal gradient and other sources of information. Mineral physicists study the elastic properties of minerals; their high-pressure phase diagrams, melting points and equations of state at high pressure; and the rheological properties of rocks, or their ability to flow. Deformation of rocks by creep make flow possible, although over short times the rocks are brittle. The viscosity of rocks is affected by temperature and pressure, and in turn determines the rates at which tectonic plates move.

Water is a very complex substance and its unique properties are essential for life. Its physical properties shape the hydrosphere and are an essential part of the water cycle and climate. Its thermodynamic properties determine evaporation and the thermal gradient in the atmosphere. The many types of precipitation involve a complex mixture of processes such as coalescence, supercooling and supersaturation. Some precipitated water becomes groundwater, and groundwater flow includes phenomena such as percolation, while the conductivity of water makes electrical and electromagnetic methods useful for tracking groundwater flow. Physical properties of water such as salinity have a large effect on its motion in the oceans.

The many phases of ice form the cryosphere and come in forms like ice sheets, glaciers, sea ice, freshwater ice, snow, and frozen ground (or permafrost).

Regions of the Earth

Size and Form of the Earth

The Earth is roughly spherical, but it bulges towards the Equator, so it is roughly in the shape of an ellipsoid. This bulge is due to its rotation and is nearly consistent with an Earth in hydrostatic equilibrium. The detailed shape of the Earth, however, is also affected by the distribution of continents and ocean basins, and to some extent by the dynamics of the plates.

Structure of the Interior

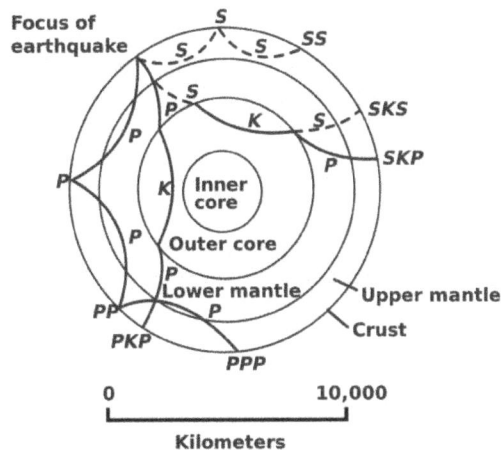

Seismic velocities and boundaries in the interior of the Earth sampled by seismic waves.

Evidence from seismology, heat flow at the surface, and mineral physics is combined with the Earth's mass and moment of inertia to infer models of the Earth's interior – its composition, density, temperature, pressure. For example, the Earth's mean specific gravity (5.515) is far higher than the typical specific gravity of rocks at the surface (2.7–3.3), implying that the deeper material is denser. This is also implied by its low moment of inertia ($0.33\,M\,R^2$, compared to $0.4\,M\,R^2$ for a sphere of constant density). However, some of the density increase is compression under the enormous pressures inside the Earth. The effect of pressure can be calculated using the Adams–Williamson equation. The conclusion is that pressure alone cannot account for the increase in density. Instead, we know that the Earth's core is composed of an alloy of iron and other minerals.

Reconstructions of seismic waves in the deep interior of the Earth show that there are no S-waves in the outer core. This indicates that the outer core is liquid, because liquids cannot support shear. The outer core is liquid, and the motion of this highly conductive fluid generates the Earth's field. The inner core, however, is solid because of the enormous pressure.

Reconstruction of seismic reflections in the deep interior indicate some major discontinuities in seismic velocities that demarcate the major zones of the Earth: inner core, outer core, mantle, lithosphere and crust. The mantle itself is divided into the upper mantle, transition zone, lower mantle and D'' layer. Between the crust and the mantle is the Mohorovičić discontinuity.

The seismic model of the Earth does not by itself determine the composition of the layers. For a complete model of the Earth, mineral physics is needed to interpret seismic velocities in terms of composition. The mineral properties are temperature-dependent, so the geotherm must also be determined. This requires physical theory for thermal conduction and convection and the heat contribution of radioactive elements. The main model for the radial structure of the interior of the Earth is the preliminary reference Earth model (PREM). Some parts of this model have been updated by recent findings in mineral physics and supplemented by seismic tomography. The mantle is mainly composed of silicates, and the boundaries between layers of the mantle are consistent with phase transitions.

The mantle acts as a solid for seismic waves, but under high pressures and temperatures it deforms so that over millions of years it acts like a liquid. This makes plate tectonics possible. Geodynamics is the study of the fluid flow in the mantle and core.

Magnetosphere

If a planet's magnetic field is strong enough, its interaction with the solar wind forms a magnetosphere. Early space probes mapped out the gross dimensions of the Earth's magnetic field, which extends about 10 Earth radii towards the Sun. The solar wind, a stream of charged particles, streams out and around the terrestrial magnetic field, and continues behind the magnetic tail, hundreds of Earth radii downstream. Inside the magnetosphere, there are relatively dense regions of solar wind particles called the Van Allen radiation belts.

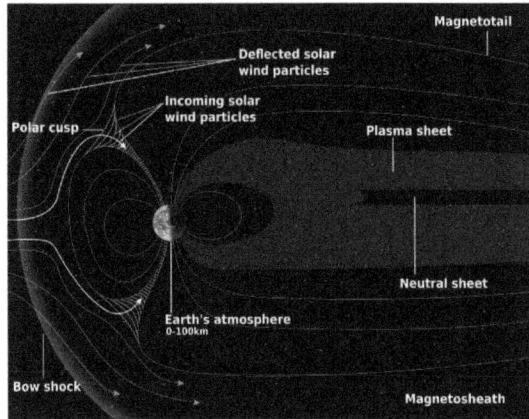

Schematic of Earth's magnetosphere. The solar wind flows from left to right.

Methods

Geodesy

Geophysical measurements are generally at a particular time and place. Accurate measurements of position, along with earth deformation and gravity, are the province of geodesy. While geodesy and geophysics are separate fields, the two are so closely connected that many scientific organizations such as the American Geophysical Union, the Canadian Geophysical Union and the International Union of Geodesy and Geophysics encompass both.

Absolute positions are most frequently determined using the global positioning system (GPS). A three-dimensional position is calculated using messages from four or more visible satellites and referred to the 1980 Geodetic Reference System. An alternative, optical astronomy, combines astronomical coordinates and the local gravity vector to get geodetic coordinates. This method only provides the position in two coordinates and is more difficult to use than GPS. However, it is useful for measuring motions of the Earth such as nutation and Chandler wobble. Relative positions of two or more points can be determined using very-long-baseline interferometry.

Gravity measurements became part of geodesy because they were needed to related measurements at the surface of the Earth to the reference coordinate system. Gravity measurements on land can be made using gravimeters deployed either on the surface or in helicopter flyovers. Since the 1960s, the Earth's gravity field has been measured by analyzing the motion of satellites. Sea level can also be measured by satellites using radar altimetry, contributing to a more accurate geoid. In 2002, NASA launched the Gravity Recovery and Climate Experiment (GRACE), wherein two twin satellites map variations in Earth's gravity field by making measurements of the distance between the two satellites using GPS and a microwave ranging system. Gravity variations detected by GRACE include those caused by changes in ocean currents; runoff and ground water depletion; melting ice sheets and glaciers.

Space Probes

Space probes made it possible to collect data from not only the visible light region, but in other areas of the electromagnetic spectrum. The planets can be characterized by their force fields: gravity and their magnetic fields, which are studied through geophysics and space physics.

Measuring the changes in acceleration experienced by spacecraft as they orbit has allowed fine details of the gravity fields of the planets to be mapped. For example, in the 1970s, the gravity field disturbances above lunar maria were measured through lunar orbiters, which led to the discovery of concentrations of mass, mascons, beneath the Imbrium, Serenitatis, Crisium, Nectaris and Humorum basins.

History

Geophysics emerged as a separate discipline only in the 19th century, from the intersection of physical geography, geology, astronomy, meteorology, and physics. However, many geophysical phenomena – such as the Earth's magnetic field and earthquakes – have been investigated since the ancient era.

Ancient and Classical Eras

The magnetic compass existed in China back as far as the fourth century BC. It was used as much for feng shui as for navigation on land. It was not until good steel needles could be forged that compasses were used for navigation at sea; before that, they could not retain their magnetism long enough to be useful. The first mention of a compass in Europe was in 1190 AD.

Replica of Zhang Heng's seismoscope, possibly the first contribution to seismology.

In circa 240 BC, Eratosthenes of Cyrene deduced that the Earth was round and measured the circumference of the Earth, using trigonometry and the angle of the Sun at more than one latitude in Egypt. He developed a system of latitude and longitude.

Perhaps the earliest contribution to seismology was the invention of a seismoscope by the prolific inventor Zhang Heng in 132 AD. This instrument was designed to drop a bronze ball from the mouth of a dragon into the mouth of a toad. By looking at which of eight toads had the ball, one could determine the direction of the earthquake. It was 1571 years before the first design for a seismoscope was published in Europe, by Jean de la Hautefeuille. It was never built.

Beginnings of Modern Science

One of the publications that marked the beginning of modern science was William Gilbert's *De Magnete* (1600), a report of a series of meticulous experiments in magnetism. Gilbert deduced that compasses point north because the Earth itself is magnetic.

In 1687 Isaac Newton published his *Principia*, which not only laid the foundations for classical mechanics and gravitation but also explained a variety of geophysical phenomena such as the tides and the precession of the equinox.

The first seismometer, an instrument capable of keeping a continuous record of seismic activity, was built by James Forbes in 1844.

Historical Geology

Historical geology is a discipline that uses the principles and techniques of geology to reconstruct and understand the geological history of Earth. It focuses on geologic processes that change the Earth's surface and subsurface; and the use of stratigraphy, structural geology and paleontology to tell the sequence of these events. It also focuses on the evolution of plants and animals during different time periods in the geological timescale. The discovery of radioactivity and the development of a variety of radiometric dating techniques in the first half of the 20th century provided a means of deriving absolute versus relative ages of geologic history.

Economic geology, the search for and extraction of energy and raw materials, is heavily dependent on an understanding of the geological history of an area. Environmental geology, including most importantly the geologic hazards of earthquakes and volcanism, must also include a detailed knowledge of geologic history.

Historical Development

Nicolaus Steno, also known as Niels Stensen, was the first to observe and propose some of the basic concepts of historical geology. One of these concepts was that fossils originally came from living organisms. The other, more famous, observations are often grouped together to form the laws of stratigraphy.

James Hutton and Charles Lyell also contributed to early understanding of the Earth's history with their observations at Edinburgh in Scotland concerning angular unconformity in a rock face and it was in fact Lyell that influenced Charles Darwin greatly in his theory of evolution by speculating that *the present is the key to the past*. Hutton first proposed the theory of uniformitarianism, which is now a basic principle in all branches of geology. Hutton also supported the idea that the Earth was very old as opposed to the prevailing concept of the time which said the Earth had only been around a few millennia. Uniformitarianism describes an Earth created by the same natural phenomena that are at work today.

The prevailing concept of the 18th century in the West was that of a very short Earth history dominated by catastrophic events. This view was strongly supported by adherents of Abrahamic religions based on a largely literal interpretation of their religious scriptural passages. The concept of uniformitarianism met with considerable resistance and the catastrophism vs. gradualism debate of the 19th century resulted. A variety of discoveries in the 20th century provided ample evidence that Earth history is a product of both gradual incremental processes and sudden cataclysmic events. Violent events such as meteorite impacts and large volcanic explosions do shape the Earth's surface along with gradual processes such as weathering, erosion and deposition much as they have throughout Earth history. *The present is the key to the past* - includes catastrophic as well as gradual processes.

Hydrogeology

Hydrogeology (*hydro-* meaning water, and *-geology* meaning the study of the Earth) is the area of geology that deals with the distribution and movement of groundwater in the soil and rocks of the Earth's crust (commonly in aquifers). The term geohydrology is often used interchangeably. Some make the minor distinction between a hydrologist or engineer applying themselves to geology (geohydrology), and a geologist applying themselves to hydrology (hydrogeology).

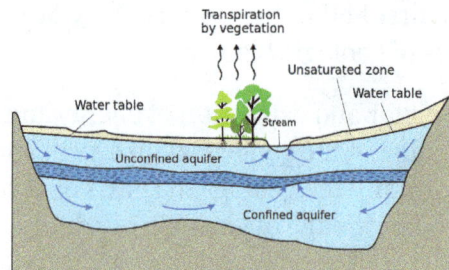

Typical aquifer cross-section

Introduction

Hydrogeology is an interdisciplinary subject; it can be difficult to account fully for the chemical, physical, biological and even legal interactions between soil, water, nature and society. The study of the interaction between groundwater movement and geology can be quite complex. Groundwater does not always flow in the sub-surface down-hill following the surface topography; groundwater follows pressure gradients (flow from high pressure to low) often following fractures and conduits in circuitous paths. Taking into account the interplay of the different facets of a multi-component system often requires knowledge in several diverse fields at both the experimental and theoretical levels. The following is a more traditional intro-duction to the methods and nomenclature of saturated subsurface hydrology, or simply the study of ground water content.

Hydrogeology in Relation to Other Fields

Painting of Ivan Aivazovsky (1841).

Hydrogeology, as stated above, is a branch of the earth sciences dealing with the flow of water through aquifers and other shallow porous media (typically less than 450 m or 1,500 ft below the land surface.) The very shallow flow of water in the subsurface (the upper 3 m or 10 ft) is pertinent to the fields of soil science, agriculture and civil engi-neering, as well as to hydrogeology. The general flow of fluids (water, hydrocarbons, geothermal fluids, etc.) in deeper formations is also a concern of geologists, geophys-icists and petroleum geologists. Groundwater is a slow-moving, viscous fluid (with a Reynolds number less than unity); many of the empirically derived laws of groundwa-ter flow can be alternately derived in fluid mechanics from the special case of Stokes flow (viscosity and pressure terms, but no inertial term).

The mathematical relationships used to describe the flow of water through porous me-dia are the diffusion and Laplace equations, which have applications in many diverse fields. Steady groundwater flow (Laplace equation) has been simulated using electrical, elastic and heat conduction analogies. Transient groundwater flow is analogous to the diffusion of heat in a solid, therefore some solutions to hydrological problems have been adapted from heat transfer literature.

Traditionally, the movement of groundwater has been studied separately from surface water, climatology, and even the chemical and microbiological aspects of hydrogeology (the processes are uncoupled). As the field of hydrogeology matures, the strong inter-actions between groundwater, surface water, water chemistry, soil moisture and even climate are becoming more clear.

For example: Aquifer drawdown or overdrafting and the pumping of fossil water may be a contributing factor to sea-level rise.

Definitions and Material Properties

One of the main tasks a hydrogeologist typically performs is the prediction of future behavior of an aquifer system, based on analysis of past and present observations. Some hypothetical, but characteristic questions asked would be:

- Can the aquifer support another subdivision?

- Will the river dry up if the farmer doubles his irrigation?

- Did the chemicals from the dry cleaning facility travel through the aquifer to my well and make me sick?

- Will the plume of effluent leaving my neighbor's septic system flow to my drinking water well?

Most of these questions can be addressed through simulation of the hydrologic system (using numerical models or analytic equations). Accurate simulation of the aquifer system requires knowledge of the aquifer properties and boundary conditions. Therefore, a common task of the hydrogeologist is determining aquifer properties using aquifer tests.

In order to further characterize aquifers and aquitards some primary and derived physical properties are introduced below. Aquifers are broadly classified as being either confined or unconfined (water table aquifers), and either saturated or unsaturated; the type of aquifer affects what properties control the flow of water in that medium (e.g., the release of water from storage for confined aquifers is related to the storativity, while it is related to the specific yield for unconfined aquifers).

Hydraulic Head

Differences in hydraulic head (h) cause water to move from one place to another; water flows from locations of high h to locations of low h. Hydraulic head is composed of pressure head (ψ) and elevation head (z). The head gradient is the change in hydraulic head per length of flowpath, and appears in Darcy's law as being proportional to the discharge.

Hydraulic head is a directly measurable property that can take on any value (because of the arbitrary datum involved in the z term); ψ can be measured with a pressure transducer (this value can be negative, e.g., suction, but is positive in saturated aquifers), and z can be measured relative to a surveyed datum (typically the top of the well casing). Commonly, in wells tapping unconfined aquifers the water level in a well is used as a proxy for hydraulic head, assuming there is no vertical gradient of pressure. Often only *changes* in hydraulic head through time are needed, so the constant elevation head term can be left out ($\Delta h = \Delta \psi$).

A record of hydraulic head through time at a well is a hydrograph or, the changes in hydraulic head recorded during the pumping of a well in a test are called drawdown.

Porosity

Porosity (n) is a directly measurable aquifer property; it is a fraction between 0 and 1 indicating the amount of pore space between unconsolidated soil particles or within a fractured rock. Typically, the majority of groundwater (and anything dissolved in it) moves through the porosity available to flow (sometimes called effective porosity). Permeability is an expression of the connectedness of the pores. For instance, an unfractured rock unit may have a high *porosity* (it has lots of *holes* between its constituent grains), but a low *permeability* (none of the pores are connected). An example of this phenomenon is pumice, which, when in its unfractured state, can make a poor aquifer.

Porosity does not directly affect the distribution of hydraulic head in an aquifer, but it has a very strong effect on the migration of dissolved contaminants, since it affects groundwater flow velocities through an inversely proportional relationship.

Water Content

Water content (θ) is also a directly measurable property; it is the fraction of the total rock which is filled with liquid water. This is also a fraction between 0 and 1, but it must also be less than or equal to the total porosity.

The water content is very important in vadose zone hydrology, where the hydraulic conductivity is a strongly nonlinear function of water content; this complicates the solution of the unsaturated groundwater flow equation.

Hydraulic Conductivity

Hydraulic conductivity (K) and transmissivity (T) are indirect aquifer properties (they cannot be measured directly). T is the K integrated over the vertical thickness (b) of the aquifer ($T=Kb$ when K is constant over the entire thickness). These properties are measures of an aquifer's ability to transmit water. Intrinsic permeability (κ) is a secondary medium property which does not depend on the viscosity and density of the fluid (K and T are specific to water); it is used more in the petroleum industry.

Specific Storage and Specific Yield

Specific storage (S_s) and its depth-integrated equivalent, storativity ($S=S_s b$), are indirect aquifer properties (they cannot be measured directly); they indicate the amount of groundwater released from storage due to a unit depressurization of a confined aquifer. They are fractions between 0 and 1.

Specific yield (S_y) is also a ratio between 0 and 1 ($S_y \leq$ porosity) and indicates the amount of water released due to drainage from lowering the water table in an unconfined aquifer. The value for specific yield is less than the value for porosity because some water will remain in the medium even after drainage due to intermolecular forces. Often the porosity or effective porosity is used as an upper bound to the specific yield. Typically S_y is orders of magnitude larger than S_s.

Contaminant Transport Properties

Often we are interested in how the moving groundwater will transport dissolved contaminants around (the sub-field of contaminant hydrogeology). The contaminants can be man-made (e.g., petroleum products, nitrate, Chromium or radionuclides) or naturally occurring (e.g., arsenic, salinity). Besides needing to understand where the groundwater is flowing, based on the other hydrologic properties discussed above, there are additional aquifer properties which affect how dissolved contaminants move with groundwater.

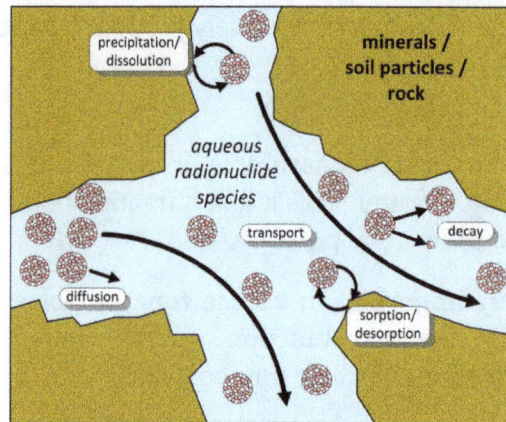

Transport and fate of contaminants in groundwater

Hydrodynamic Dispersion

Hydrodynamic dispersivity (α_L, α_T) is an empirical factor which quantifies how much contaminants stray away from the path of the groundwater which is carrying it. Some of the contaminants will be "behind" or "ahead" the mean groundwater, giving rise to a longitudinal dispersivity (α_L), and some will be "to the sides of" the pure advective groundwater flow, leading to a transverse dispersivity (α_T). Dispersion in groundwater arises because each water "particle", passing beyond a soil particle, must choose where to go, whether left or right or up or down, so that the water "particles" (and their solute) are gradually spread in all directions around the mean path. This is the "microscopic" mechanism, on the scale of soil particles. More important, on long distances, can be the macroscopic inhomogeneities of the aquifer, which can have regions of larger or smaller permeability, so that some water can find a preferential path in one direction, some other in a different direction, so that the contaminant can be spread in a completely irregular way, like in a (three-dimensional) delta of a river.

Dispersivity is actually a factor which represents our *lack of information* about the system we are simulating. There are many small details about the aquifer which are being averaged when using a macroscopic approach (e.g., tiny beds of gravel and clay in sand aquifers), they manifest themselves as an *apparent* dispersivity. Because of this, α is often claimed to be dependent on the length scale of the problem — the dispersivity found for transport through 1 m³ of aquifer is different from that for transport through 1 cm³ of the same aquifer material.

Molecular Diffusion

Diffusion is a fundamental physical phenomenon, which Einstein characterized as Brownian motion, that describes the random thermal movement of molecules and small particles in gases and liquids. It is an important phenomenon for small distances (it is essential for the achievement of thermodynamic equilibria), but, as the time necessary to cover a distance by diffusion is proportional to the square of the distance itself, it is ineffective for spreading a solute over macroscopic distances. The diffusion coefficient, D, is typically quite small, and its effect can often be considered negligible (unless groundwater flow velocities are extremely low, as they are in clay aquitards).

It is important not to confuse diffusion with dispersion, as the former is a physical phenomenon and the latter is an empirical factor which is cast into a similar form as diffusion, because we already know how to solve that problem.

Retardation by Adsorption

The retardation factor is another very important feature that make the motion of the contaminant to deviate from the average groundwater motion. It is analogous to the retardation factor of chromatography. Unlike diffusion and dispersion, which simply spread the contaminant, the retardation factor changes its *global average velocity*, so that it can be much slower than that of water. This is due to a chemico-physical effect: the adsorption to the soil, which holds the contaminant back and does not allow it to progress until the quantity corresponding to the chemical adsorption equilibrium has been adsorbed. This effect is particularly important for less soluble contaminants, which thus can move even hundreds or thousands times slower than water. The effect of this phenomenon is that only more soluble species can cover long distances. The retardation factor depends on the chemical nature of both the contaminant and the aquifer.

Governing Equations

Darcy's Law

Darcy's law is a constitutive equation, empirically derived by Henry Darcy in 1856, which states that the amount of groundwater discharging through a given portion of aquifer is proportional to the cross-sectional area of flow, the hydraulic gradient, and the hydraulic conductivity.

Groundwater Flow Equation

The groundwater flow equation, in its most general form, describes the movement of groundwater in a porous medium (aquifers and aquitards). It is known in mathematics as the diffusion equation, and has many analogs in other fields. Many solutions for groundwater flow problems were borrowed or adapted from existing heat transfer solutions.

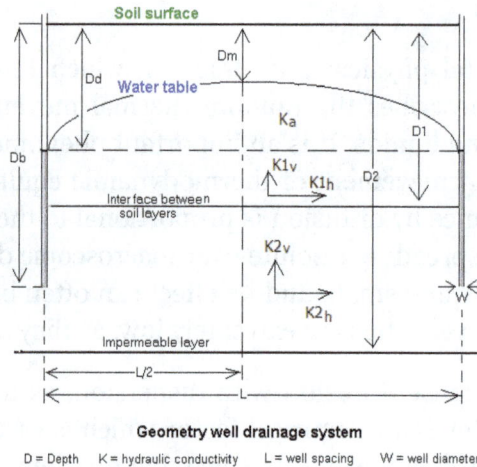

Geometry of a partially penetrating well drainage system in an anisotropic layered aquifer

It is often derived from a physical basis using Darcy's law and a conservation of mass for a small control volume. The equation is often used to predict flow to wells, which have radial symmetry, so the flow equation is commonly solved in polar or cylindrical coordinates.

The Theis equation is one of the most commonly used and fundamental solutions to the groundwater flow equation; it can be used to predict the transient evolution of head due to the effects of pumping one or a number of pumping wells.

The Thiem equation is a solution to the steady state groundwater flow equation (Laplace's Equation) for flow to a well. Unless there are large sources of water nearby (a river or lake), true steady-state is rarely achieved in reality.

Both above equations are used in aquifer tests (pump tests).

The Hooghoudt equation is a groundwater flow equation applied to subsurface drainage by pipes, tile drains or ditches. An alternative subsurface drainage method is drainage by wells for which groundwater flow equations are also available.

Calculation of Groundwater Flow

To use the groundwater flow equation to estimate the distribution of hydraulic heads, or the direction and rate of groundwater flow, this partial differential equation (PDE) must be solved. The most common means of analytically solving the diffusion equation in the hydrogeology literature are:

- Laplace, Hankel and Fourier transforms (to reduce the number of dimensions of the PDE),

- similarity transform (also called the Boltzmann transform) is commonly how the Theis solution is derived,

- separation of variables, which is more useful for non-Cartesian coordinates, and

- Green's functions, which is another common method for deriving the Theis solution — from the fundamental solution to the diffusion equation in free space.

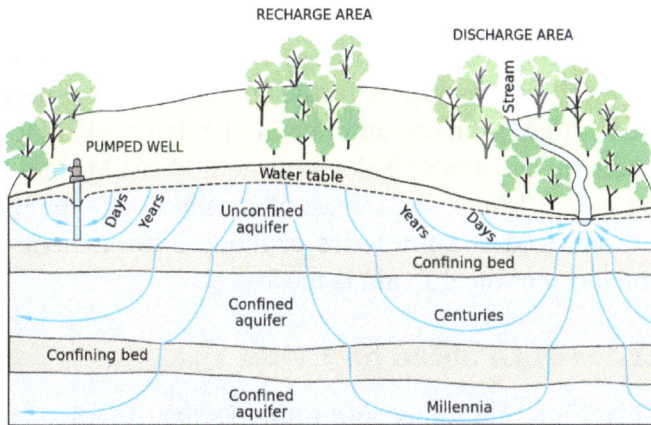

Relative groundwater travel times.

No matter which method we use to solve the groundwater flow equation, we need both initial conditions (heads at time (t) = 0) and boundary conditions (representing either the physical boundaries of the domain, or an approximation of the domain beyond that point). Often the initial conditions are supplied to a transient simulation, by a corresponding steady-state simulation (where the time derivative in the groundwater flow equation is set equal to 0).

There are two broad categories of how the (PDE) would be solved; either analytical methods, numerical methods, or something possibly in between. Typically, analytic methods solve the groundwater flow equation under a simplified set of conditions *exactly*, while numerical methods solve it under more general conditions to an *approximation*.

Analytic Methods

Analytic methods typically use the structure of mathematics to arrive at a simple, elegant solution, but the required derivation for all but the simplest domain geometries can be quite complex (involving non-standard coordinates, conformal mapping, etc.). Analytic solutions typically are also simply an equation that can give a quick answer based on a few basic parameters. The Theis equation is a very simple (yet still very useful) analytic solution to the groundwater flow equation, typically used to analyze the results of an aquifer test or slug test.

Numerical Methods

The topic of numerical methods is quite large, obviously being of use to most fields of engineering and science in general. Numerical methods have been around much longer than computers have (In the 1920s Richardson developed some of the finite difference schemes still in use today, but they were calculated by hand, using paper and pencil, by human "calculators"), but they have become very important through the availability of fast and cheap personal computers. A quick survey of the main numerical methods used in hydrogeology, and some of the most basic principles are shown below and further discussed in the Groundwater model article.

There are two broad categories of numerical methods: gridded or discretized methods and non-gridded or mesh-free methods. In the common finite difference method and finite element method (FEM) the domain is completely gridded ("cut" into a grid or mesh of small elements). The analytic element method (AEM) and the boundary integral equation method (BIEM — sometimes also called BEM, or Boundary Element Method) are only discretized at boundaries or along flow elements (line sinks, area sources, etc.), the majority of the domain is mesh-free.

General Properties of Gridded Methods

Gridded Methods like finite difference and finite element methods solve the groundwater flow equation by breaking the problem area (domain) into many small elements (squares, rectangles, triangles, blocks, tetrahedra, etc.) and solving the flow equation for each element (all material properties are assumed constant or possibly linearly variable within an element), then linking together all the elements using conservation of mass across the boundaries between the elements (similar to the divergence theorem). This results in a system which overall approximates the groundwater flow equation, but exactly matches the boundary conditions (the head or flux is specified in the elements which intersect the boundaries).

Finite differences are a way of representing continuous differential operators using discrete intervals (Δx and Δt), and the finite difference methods are based on these (they are derived from a Taylor series). For example, the first-order time derivative is often approximated using the following forward finite difference, where the subscripts indicate a discrete time location,

$$\frac{\partial h}{\partial t} = h'(t_i) \approx \frac{h_i - h_{i-1}}{\Delta t}.$$

The forward finite difference approximation is unconditionally stable, but leads to an implicit set of equations (that must be solved using matrix methods, e.g. LU or Cholesky decomposition). The similar backwards difference is only conditionally stable, but it is explicit and can be used to "march" forward in the time direction, solving one grid node at a time (or possibly in parallel, since one node depends only on its immediate neighbors). Rather than the finite difference method, sometimes the Galerkin FEM approximation is used in space (this is different from the type of FEM often used in structural engineering) with finite differences still used in time.

Application of Finite Difference Models

MODFLOW is a well-known example of a general finite difference groundwater flow model. It is developed by the US Geological Survey as a modular and extensible simulation tool for modeling groundwater flow. It is free software developed, documented and distributed by the USGS. Many commercial products have grown up around it, providing graphical user interfaces to its input file based interface, and typically incorporating pre- and post-processing of user data. Many other models have been developed to work with MODFLOW input and output, making linked models which simulate several hydrologic processes possible (flow and transport models, surface water and groundwater models and chemical reaction models), because of the simple, well documented nature of MODFLOW.

Application of Finite Element Models

Finite Element programs are more flexible in design (triangular elements vs. the block elements most finite difference models use) and there are some programs available (SUTRA, a 2D or 3D density-dependent flow model by the USGS; Hydrus, a commercial unsaturated flow model; FEFLOW, a commercial modelling environment for subsurface flow, solute and heat transport processes; OpenGeoSys, a scientific open-source project for thermo-hydro-mechanical-chemical (THMC) processes in porous and fractured media; COMSOL Multiphysics (FEMLAB) a commercial general modelling environment), FEATool Multiphysics, an easy to use Matlab simulation toolbox, and Integrated Water Flow Model (IWFM), but they are still not as popular in with practicing hydrogeologists as MODFLOW is. Finite element models are more popular in university and laboratory environments, where specialized models solve non-standard forms of the flow equation (unsaturated flow, density dependent flow, coupled heat and groundwater flow, etc.)

Application of Finite Volume Models

The finite volume method is a method for representing and evaluating partial differential equations as algebraic equations. Similar to the finite difference method, values are calculated at discrete places on a meshed geometry. "Finite volume" refers to the small volume surrounding each node point on a mesh. In the finite volume method, volume integrals in a partial differential equation that contain a divergence term are converted to surface integrals, using the divergence theorem. These terms are then evaluated as fluxes at the surfaces of each finite volume. Because the flux entering a given volume is identical to that leaving the adjacent volume, these methods are conservative.

Another advantage of the finite volume method is that it is easily formulated to allow for unstructured meshes. The method is used in many computational fluid dynamics packages.

PORFLOW software package is a comprehensive mathematical model for simulation of Ground Water Flow and Nuclear Waste Management developed by Analytic & Computational Research, Inc., ACRi.

The FEHM software package is available free from Los Alamos National Laboratory. This versatile porous flow simulator includes capabilities to model multiphase, thermal, stress, and multicomponent reactive chemistry. Current work using this code includes simulation of methane hydrate formation, CO_2 sequestration, oil shale extraction, migration of both nuclear and chemical contaminants, environmental isotope migration in the unsaturated zone, and karst formation.

Other Methods

These include mesh-free methods like the Analytic Element Method (AEM) and the Boundary Element Method (BEM), which are closer to analytic solutions, but they do approximate the groundwater flow equation in some way. The BEM and AEM exactly solve the groundwater flow equation (perfect mass balance), while approximating the boundary conditions. These methods are more exact and can be much more elegant solutions (like analytic methods are), but have not seen as widespread use outside academic and research groups yet.

Sedimentology

Sedimentology encompasses the study of modern sediments such as sand, silt, and clay, and the processes that result in their formation (erosion and weathering), transport, deposition and diagenesis. Sedimentologists apply their understanding of modern processes to interpret geologic history through observations of sedimentary rocks and sedimentary structures.

Sedimentary rocks cover up to 75% of the Earth's surface, record much of the Earth's history, and harbor the fossil record. Sedimentology is closely linked to stratigraphy, the study of the physical and temporal relationships between rock layers or strata.

The premise that the processes affecting the earth today are the same as in the past is the basis for determining how sedimentary features in the rock record were formed. By comparing similar features today to features in the rock record—for example, by comparing modern sand dunes to dunes preserved in ancient aeolian sandstones—geologists reconstruct past environments.

Sedimentary Rock Types

There are four primary types of sedimentary rocks: clastics, carbonates, evaporites, and chemical.

Middle Triassic marginal marine sequence of siltstones and sandstones, southwestern Utah.

- Clastic rocks are composed of particles derived from the weathering and erosion of precursor rocks and consist primarily of fragmental material. Clastic rocks are classified according to their predominant grain size and their composition. In the past, the term "Clastic Sedimentary Rocks" were used to describe silica-rich clastic sedimentary rocks, however there have been cases of clastic carbonate rocks. The more appropriate term is siliciclastic sedimentary rocks.

  o Organic sedimentary rocks are important deposits formed from the accumulation of biological detritus, and form coal and oil shale deposits, and are typically found within basins of clastic sedimentary rocks

- Carbonates are composed of various carbonate minerals (most often calcium carbonate ($CaCO_3$)) precipitated by a variety of organic and inorganic processes. Typically, the majority of carbonate rocks are composed of reef material.

- Evaporites are formed through the evaporation of water at the Earth's surface and most commonly include halite or gypsum.

- Chemical sedimentary rocks, including some carbonates, are deposited by precipitation of minerals from aqueous solution. These include jaspilite and chert.

Importance of Sedimentary Rocks

Sedimentary rocks provide a multitude of products which modern and ancient society has come to utilise.

Mi Vida uranium mine in redox mudstones near Moab, Utah

- Art: marble, although a metamorphosed limestone, is an example of the use of sedimentary rocks in the pursuit of aesthetics and art

- Architectural uses: stone derived from sedimentary rocks is used for dimension stone and in architecture, notably slate, a meta-shale, for roofing, sandstone for load-bearing buttresses

- Ceramics and industrial materials: clay for pottery and ceramics including bricks; cement and lime derived from limestone.

- Economic geology: sedimentary rocks host large deposits of SEDEX ore deposits of lead-zinc-silver, large deposits of copper, deposits of gold, tungsten, Uranium, and many other precious minerals, gemstones and industrial minerals including heavy mineral sands ore deposits

- Energy: petroleum geology relies on the capacity of sedimentary rocks to generate deposits of petroleum oils. Coal and oil shale are found in sedimentary rocks. A large proportion of the world's uranium energy resources are hosted within sedimentary successions.

- Groundwater: sedimentary rocks contain a large proportion of the Earth's groundwater aquifers. Our understanding of the extent of these aquifers and how much water can be withdrawn from them depends critically on our knowledge of the rocks that hold them (the reservoir).

Basic Principles

The aim of sedimentology, studying sediments, is to derive information on the depositional conditions which acted to deposit the rock unit, and the relation of the individual rock units in a basin into a coherent understanding of the evolution of the sedimentary sequences and basins, and thus, the Earth's geological history as a whole.

The scientific basis of this is the principle of uniformitarianism, which states that the sediments within ancient sedimentary rocks were deposited in the same way as sediments which are being deposited at the Earth's surface today.

Heavy minerals (dark) deposited in a quartz beach sand (Chennai, India).

Sedimentological conditions are recorded within the sediments as they are laid down; the form of the sediments at present reflects the events of the past and all events which affect the sediments, from the source of the sedimentary material to the stresses enacted upon them after diagenesis are available for study.

The principle of superposition is critical to the interpretation of sedimentary sequences, and in older metamorphic terrains or fold and thrust belts where sediments are often intensely folded or deformed, recognising younging indicators or graded bedding is critical to interpretation of the sedimentary section and often the deformation and metamorphic structure of the region.

Folding in sediments is analysed with the principle of original horizontality, which states that sediments are deposited at their angle of repose which, for most types of sediment, is essentially horizontal. Thus, when the younging direction is known, the rocks can be "unfolded" and interpreted according to the contained sedimentary information.

The principle of lateral continuity states that layers of sediment initially extend laterally in all directions unless obstructed by a physical object or topography.

The principle of cross-cutting relationships states that whatever cuts across or intrudes into the layers of strata is younger than the layers of strata.

Methodology of Sedimentology

Centripetal desiccation cracks (with a dinosaur footprint in the center) in the Lower Jurassic Moenave Formation at the St. George Dinosaur Discovery Site at Johnson Farm, southwestern Utah.

The methods employed by sedimentologists to gather data and evidence on the nature and depositional conditions of sedimentary rocks include;

- Measuring and describing the outcrop and distribution of the rock unit;

    o Describing the rock formation, a formal process of documenting thickness, lithology, outcrop, distribution, contact relationships to other formations

    o Mapping the distribution of the rock unit, or units

- Descriptions of rock core (drilled and extracted from wells during hydrocarbon exploration)

- Sequence stratigraphy

    o Describes the progression of rock units within a basin

- Describing the lithology of the rock;

    o Petrology and petrography; particularly measurement of texture, grain size, grain shape (sphericity, rounding, etc.), sorting and composition of the sediment

- Analysing the geochemistry of the rock

    o Isotope geochemistry, including use of radiometric dating, to determine the age of the rock, and its affinity to source regions

Recent Developments in Sedimentology

The longstanding understanding of how some mudstones form has been challenged by geologists at Indiana University (Bloomington) and the Massachusetts Institute of Technology. The research, which appears in the December 14th, 2007, edition of *Science*, counters the prevailing view of geologists that mud only settles when water is slow-moving or still, instead showing that "muds will accumulate even when currents move swiftly." The research shows that some mudstones may have formed in fast-moving waters: "Mudstones can be deposited under more energetic conditions than widely assumed, requiring a reappraisal of many geologic records."

Macquaker and Bohacs, in reviewing the research of Schieber et al., state that "these results call for critical reappraisal of all mudstones previously interpreted as having been continuously deposited under still waters. Such rocks are widely used to infer past climates, ocean conditions, and orbital variations."

Considerable recent research into mudstones has been driven by the recent effort to commercially produce hydrocarbons from them, in both the Shale gas and Tight Oil (or Light Tight Oil) plays.

Planetary Geology

Planetary geology, alternatively known as astrogeology or exogeology, is a planetary science discipline concerned with the geology of the celestial bodies such as the planets and their moons, asteroids, comets, and meteorites. Although the geo- prefix typically indicates topics of or relating to the Earth, planetary geology is named as such for historical and convenience reasons; applying geological science to other planetary bodies. Due to the types of investigations involved, it is also closely linked with Earth-based geology.

Planetary geologist and NASA astronaut Harrison "Jack" Schmitt collecting lunar samples during the Apollo 17 mission in early-December 1972

Surface of Mars as photographed by the Viking 2 lander December 9, 1977.

Planetary geology includes such topics as determining the internal structure of the terrestrial planets, and also looks at planetary volcanism and surface processes such as impact craters, fluvial and aeolian processes. The structures of the giant planets and their moons are also examined, as is the make-up of the minor bodies of the Solar System, such as asteroids, the Kuiper Belt, and comets.

History of Planetary Geology

Eugene Shoemaker is credited with bringing geologic principles to planetary mapping and creating the branch of planetary science in the early 1960s, the Astrogeology Research Program, within the United States Geological Survey. He made important contributions to the field and the study of impact craters, Selenography (study of the Moon), asteroids, and comets.

Today many institutions are concerned with the study and communication of planetary sciences and planetary geology. The Visitor Center at Barringer Meteor Crater near Winslow, Arizona includes a Museum of planetary geology. The Geological Society of America's Planetary Geology Division has been growing and thriving since May 1981 and has two mottos: "One planet just isn't enough!" and ""The GSA Division with the biggest field area!"

Major centers for planetary science research include the Lunar and Planetary Institute, the Applied Physics Laboratory, the Planetary Science Institute, the Jet Propulsion Laboratory, Southwest Research Institute, and Johnson Space Center. Additionally, several universities conduct extensive planetary science research, including Montana State University, Brown University, the University of Arizona, Caltech, the University of Colorado, Western Michigan University, MIT, and Washington University in St. Louis.

Features and Terms

Planetary geology uses a wide variety of standardised descriptor names for features. All planetary feature names recognised by the International Astronomical Union combine one of these names with a possibly unique identifying name. The conventions which decide the more precise name are dependent on which planetary body the feature is on, but the standard descriptors are in general common to all astronomical planetary bodies. Some names have a long history of historical usage, but new must be recognised by the IAU Working Group for Planetary System Nomenclature as features are mapped and described by new planetary missions. This means that in some cases names may change as new imagery becomes available, or in other cases widely adopted informal names changed in line with the rules. The standard names are chosen to consciously avoid interpreting the underlying cause of the feature, but rather to describe only its appearance.

Feature	Pronunciation	Description	Designation
Albedo feature	/ælˈbiːdoʊ/	An area which shows a contrast in brightness or darkness (albedo) with adjacent areas. This term is implicit.	AL
Arcus, arcūs	/ˈɑːrkəs/	Arc: curved feature	AR
Astrum, astra	/ˈæstrəm/, /ˈæstrə/	Radial-patterned features on Venus	AS
Catena, catenae	/kəˈtiːnə/, /kəˈtiːniː/	A chain of craters e.g. Enki Catena.	CA

Cavus, cavi	/ˈkeɪvəs/, /ˈkeɪvaɪ/	Hollows, irregular steep-sided depressions usually in arrays or clusters	CB
Chaos	/ˈkeɪ.ɒs/	A distinctive area of broken or jumbled terrain e.g. Iani Chaos.	CH
Chasma, chasmata	/ˈkæzmə/, /ˈkæzmətə/	Deep, elongated, steep-sided depression e.g. Eos Chasma.	CM
Colles	/ˈkɒliːz/	A collection of small hills or knobs.	CO
Corona, coronae	/kɒˈroʊnə/, /kɒˈroʊniː/	An oval feature. Used only on Venus and Miranda.	CR
Crater, craters	/ˈkreɪtər/	A circular depression likely created by impact event. This term is implicit.	AA
Dorsum, dorsa	/ˈdɔːrsəm/, /ˈdɔːrsə/	Ridge, sometimes called a wrinkle ridge e.g. Dorsum Buckland.	DO
Eruptive center		An active volcano on Io. This term is implicit.	ER
Facula, faculae	/ˈfækjʊlə/, /ˈfækjʊliː/	Bright spot	FA
Farrum, farra	/ˈfærəm/, /ˈfærə/	Pancake-like structure, or a row of such structures. Used only on Venus.	FR
Flexus, flexūs	/ˈflɛksəs/	Very low curvilinear ridge with a scalloped pattern	FE
Fluctus, fluctūs	/ˈflʌktəs/	Terrain covered by outflow of liquid. Used on Venus, Io and Titan.	FL
Flumen, flumina	/ˈfluːmɪn/, /ˈfluːmɪnə/	Channel on Titan that might carry liquid	FM
Fossa, fossae	/ˈfɒsə/, /ˈfɒsiː/	Long, narrow, shallow depression	FO
Fretum, freta	/ˈfriːtəm/, /ˈfriːtə/	Strait of liquid connecting two larger areas of liquid. Used only on Titan.	FT
Insula, insulae	/ˈɪnsjuːlə/, /ˈɪnsjuːliː/	Island (islands), an isolated land area (or group of such areas) surrounded by, or nearly surrounded by, a liquid area (sea or lake). Used only on Titan.	IN
Labes, labes	/ˈleɪbiːz/	Landslide debris. Used only on Mars.	LA
Labyrinthus, labyrinthi	/læbɪˈrɪnθəs/, /læbɪˈrɪnθaɪ/	Complex of intersecting valleys or ridges.	LB
Lacuna, lacunae	/ləˈkjuːnə/, /ləˈkjuːniː/	Irregularly shaped depression having the appearance of a dry lake bed. Used only on Titan.	LU
Lacus, lacūs	/ˈleɪkəs/	A "lake" or small plain on Moon and Mars; on Titan, a "lake" or small, dark plain with discrete, sharp boundaries.	LC
Landing site name		Lunar features at or near Apollo landing sites	LF
Large ringed feature		Cryptic ringed features	LG

Lenticula, lenticulae	/lɛnˈtɪkjʊlə/, /lɛnˈtɪkjʊliː/	Small dark spots on Europa	LE
Linea, lineae	/ˈlɪniːə/, /ˈlɪniːˌiː/	Dark or bright elongate marking, may be curved or straight	LI
Macula, maculae	/ˈmækjʊlə/, /ˈmækjʊliː/	Dark spot, may be irregular	MA
Mare, maria	/ˈmɑːriː/ ~ /ˈmɑːreɪ/, /ˈmɑːriə/	A "sea" or large circular plain on Moon and Mars, e.g. Mare Erythraeum; on Titan, large expanses of dark materials thought to be liquid hydrocarbons, e.g. Ligeia Mare.	ME
Mensa, mensae	/ˈmɛnsə/, /ˈmɛnsiː/	A flat-topped prominence with cliff-like edges, i.e. a mesa.	MN
Mons, montes	/ˈmɒnz/, /ˈmɒntiːz/	Mons refers to a mountain. Montes refers to a mountain range.	MO
Oceanus	/oʊʃiˈaɪnəs/	Very large dark area. The only feature with this designation is Oceanus Procellarum.	OC
Palus, paludes	/ˈpeɪləs/, /pəˈljuːdiːz/	"Swamp"; small plain. Used on the Moon and Mars.	PA
Patera, paterae	/ˈpætərə/, /ˈpætəriː/	Irregular crater, or a complex one with scalloped edges e.g. Ah Peku Patera. Usually refers to the dish-shaped depression atop a volcano.	PE
Planitia, planitiae	/pləˈnɪʃə/, /pləˈnɪʃiː/	Low plain e.g. Amazonis Planitia.	PL
Planum, plana	/ˈpleɪnəm/, /ˈpleɪnə/	A plateau or high plain e.g. Planum Boreum.	PM
Plume		A cryovolcanic feature on Triton. This term is currently unused.	PU
Promontorium, promontoria	/prɒmənˈtɔəriəm/, /prɒmənˈtɔəriə/	"Cape"; headland. Used only on the Moon.	PR
Regio, regiones	/ˈriːdʒioʊ/ ~ /ˈrɛdʒioʊ/, /rɛdʒiˈoʊniːz/	Large area marked by reflectivity or color distinctions from adjacent areas, or a broad geographic region	RE
Reticulum, reticula	/rɪˈtɪkjʊləm/, /rɪˈtɪkjʊlə/	reticular (netlike) pattern on Venus	RT
Rima, rimae	/ˈraɪmə/, /ˈraɪmiː/	Fissure. Used only on the Moon.	RI
Rupes, rupes	/ˈruːpiːz/	Scarp	RU
Satellite feature		A feature that shares the name of an associated feature, for example Hertzsprung D.	SF
Scopulus, scopuli	/ˈskɒpjʊlə/, /ˈskɒpjʊlaɪ/	Lobate or irregular scarp	SC
Serpens, serpentes	/ˈsɜːrpɛnz/, /sərˈpɛntiːz/	Sinuous feature with segments of positive and negative relief along its length	SE

Sinus	/ˈsaɪnəs/	"Bay"; small plain on Moon or Mars, e.g. Sinus Meridiani; On Titan, bay within bodies of liquid.	SI
Sulcus, sulci	/ˈsʌlkəs/, /ˈsʌlsaɪ/	Subparallel furrows and ridges	SU
Terra, terrae	/ˈtɛrə/, /ˈtɛriː/	Extensive land mass e.g. Arabia Terra, Aphrodite Terra.	TA
Tessera, tesserae	/ˈtɛsərə/, /ˈtɛsəriː/	An area of tile-like, polygonal terrain. This term is used only on Venus.	TE
Tholus, tholi	/ˈθoʊləs/, /ˈθoʊlaɪ/	Small domical mountain or hill e.g. Hecates Tholus.	TH
Undae	/ˈʌndiː/	A field of dunes. Used on Venus, Mars and Titan.	UN
Vallis, valles	/ˈvælɪs/, /ˈvæliːz/	A valley e.g. Valles Marineris.	VA
Vastitas, vastitates	/ˈvæstɪtəs/, /væstɪˈteɪtiːz/	An extensive plain. The only feature with this designation is Vastitas Borealis.	VS
Virga, virgae	/ˈvɜːrgə/, /ˈvɜːrdʒiː/	A streak or stripe of color. This term is currently used only on Titan.	VI

Geochemistry

Geochemistry is the science that uses the tools and principles of chemistry to explain the mechanisms behind major geological systems such as the Earth's crust and its oceans. The realm of geochemistry extends beyond the Earth, encompassing the entire Solar System and has made important contributions to the understanding of a number of processes including mantle convection, the formation of planets and the origins of granite and basalt.

History

The term *geochemistry* was first used by the Swiss-German chemist Christian Friedrich Schönbein in 1838. In his paper, Schönbein predicted the birth of a new field of study, stating:

"In a word, a comparative geochemistry ought to be launched, before geochemistry can become geology, and before the mystery of the genesis of our planets and their inorganic matter may be revealed."

The field began to be realised a short time after Schönbein's work, but his term - 'geochemistry' - was initially used neither by geologists nor chemists and there was much debate over which of the two sciences should be the dominant partner. There was little collaboration between geologists and chemists and the field of geochemistry remained small and unrecognised. In the late 19th Century a Swiss man by the name of Victor

Goldschmidt was born, who later became known as the father of geochemistry. His paper, Geochemische Verteilungsgesetze der Elemente, on the distribution of elements in nature has been referred to as the start of geochemistry. During the early 20th Century, a number of geochemists produced work that began to popularise the field, including Frank Wigglesworth Clarke who had begun to investigate the abundances of various elements within the Earth and how the quantities were related to atomic weight. The composition of meteorites and their differences to terrestrial rocks was being investigated as early as 1850 and in 1901, Oliver C. Farrington hypothesised although there were differences, that the relative abundances should still be the same. This was the beginnings of the field of cosmochemistry and has contributed much of what we know about the formation of the Earth and the Solar System.

Subfields

Some subsets of geochemistry are:

1. Isotope geochemistry involves the determination of the relative and absolute concentrations of the elements and their isotopes in the earth and on earth's surface.

2. Examination of the distribution and movements of elements in different parts of the earth (crust, mantle, hydrosphere etc.) and in minerals with the goal to determine the underlying system of distribution and movement.

3. Cosmochemistry includes the analysis of the distribution of elements and their isotopes in the cosmos.

4. Biogeochemistry is the field of study focusing on the effect of life on the chemistry of the earth.

5. Organic geochemistry involves the study of the role of processes and compounds that are derived from living or once-living organisms.

6. Aqueous geochemistry studies the role of various elements in watersheds, including copper, sulfur, mercury, and how elemental fluxes are exchanged through atmospheric-terrestrial-aquatic interactions.

7. Regional, environmental and exploration geochemistry includes applications to environmental, hydrological and mineral exploration studies.

8. Photogeochemistry is the study of light-induced chemical reactions that occur or may occur among natural components of the earth's surface.

Victor Goldschmidt is considered by most to be the father of modern geochemistry and the ideas of the subject were formed by him in a series of publications from 1922 under the title 'Geochemische Verteilungsgesetze der Elemente' (geochemical laws of distribution of the elements).

Chemical Characteristics

The more common rock constituents are nearly all oxides; chlorides, sulfides and fluorides are the only important exceptions to this and their total amount in any rock is usually much less than 1%. F. W. Clarke has calculated that a little more than 47% of the Earth's crust consists of oxygen. It occurs principally in combination as oxides, of which the chief are silica, alumina, iron oxides, and various carbonates (calcium carbonate, magnesium carbonate, sodium carbonate, and potassium carbonate). The silica functions principally as an acid, forming silicates, and all the commonest minerals of igneous rocks are of this nature. From a computation based on 1672 analyses of numerous kinds of rocks Clarke arrived at the following as the average percentage composition of the earths crust: SiO_2=59.71, Al_2O_3=15.41, Fe_2O_3=2.63, FeO=3.52, MgO=4.36, CaO=4.90, Na_2O=3.55, K_2O=2.80, H_2O=1.52, TiO_2=0.60, P_2O_5=0.22, (total 99.22%). All the other constituents occur only in very small quantities, usually much less than 1%.

These oxides combine in a haphazard way. For example, potash (potassium carbonate) and soda (sodium carbonate) combine to produce feldspars. In some cases they may take other forms, such as nepheline, leucite, and muscovite, but in the great majority of instances they are found as feldspar. Phosphoric acid with lime (calcium carbonate) forms apatite. Titanium dioxide with ferrous oxide gives rise to ilmenite. Part of the lime forms lime feldspar. Magnesium carbonate and iron oxides with silica crystallize as olivine or enstatite, or with alumina and lime form the complex ferro-magnesian silicates of which the pyroxenes, amphiboles, and biotites are the chief. Any excess of silica above what is required to neutralize the bases will separate out as quartz; excess of alumina crystallizes as corundum. These must be regarded only as general tendencies. It is possible, by rock analysis, to say approximately what minerals the rock contains, but there are numerous exceptions to any rule.

Mineral Constitution

Hence we may say that except in acid or siliceous rocks containing greater than 66% of silica are known as felsic rocks, and Quartz is not abundant. In basic rocks (containing 20% of silica or less) it is rare for them to contain as much silicon, these are referred to as mafic rocks. If Magnesium and Iron are above average while silica is low, olivine may be expected; where silica is present in greater quantity over ferro-magnesian minerals, such as augite, hornblende, enstatite or biotite, occur rather than olivine. Unless potash is high and silica relatively low, leucite will not be present, for leucite does not occur with free quartz. Nepheline, likewise, is usually found in rocks with much soda and comparatively little silica. With high alkalis, soda-bearing pyroxenes and amphiboles may be present. The lower the percentage of silica and alkali's, the greater is the prevalence of plagioclase feldspar as contracted with soda or potash feldspar. The earth crust is composed of 90% silicate minerals and their abundance in the earth is as follows; plagioclase feldspar (39%), Alkali feldspar

(12%), quartz (12%), pyroxene (11%), amphiboles (5%), micas (5%), clay minerals (5%), after this the remaining silicate minerals make up another 3% of the earths crust. Only 8% of the earth is composed of non silicate minerals such as Carbonate, Oxides, and Sulfides.

The other determining factor, namely the physical conditions attending consolidation, plays on the whole a smaller part, yet is by no means negligible, as a few instances will prove. Certain minerals are practically confined to deep-seated intrusive rocks, e.g., microcline, muscovite, diallage. Leucite is very rare in plutonic masses; many minerals have special peculiarities in microscopic character according to whether they crystallized in depth or near the surface, e.g., hypersthene, orthoclase, quartz. There are some curious instances of rocks having the same chemical composition, but consisting of entirely different minerals, e.g., the hornblendite of Gran, in Norway, which contains only hornblende, has the same composition as some of the camptonites of the same locality that contain feldspar and hornblende of a different variety. In this connection we may repeat what has been said above about the corrosion of porphyritic minerals in igneous rocks. In rhyolites and trachytes, early crystals of hornblende and biotite may be found in great numbers partially converted into augite and magnetite. Hornblende and biotite were stable under the pressures and other conditions below the surface, but unstable at higher levels. In the ground-mass of these rocks, augite is almost universally present. But the plutonic representatives of the same magma, granite and syenite contain biotite and hornblende far more commonly than augite.

Felsic, Intermediate and Mafic Igneous Rocks

Those rocks that contain the most silica, and on crystallizing yield free quartz, form a group generally designated the "felsic" rocks. Those again that contain least silica and most magnesia and iron, so that quartz is absent while olivine is usually abundant, form the "mafic" group. The "intermediate" rocks include those characterized by the general absence of both quartz and olivine. An important subdivision of these contains a very high percentage of alkalis, especially soda, and consequently has minerals such as nepheline and leucite not common in other rocks. It is often separated from the others as the "alkali" or "soda" rocks, and there is a corresponding series of mafic rocks. Lastly a small sub-group rich in olivine and without feldspar has been called the "ultramafic" rocks. They have very low percentages of silica but much iron and magnesia.

Except these last, practically all rocks contain felspars or feldspathoid minerals. In the acid rocks the common feldspars are orthoclase, perthite, microcline, and oligoclase—all having much silica and alkalis. In the mafic rocks labradorite, anorthite and bytownite prevail, being rich in lime and poor in silica, potash and soda. Augite is the most common ferro-magnesian in mafic rocks, but biotite and hornblende are on the whole more frequent in felsic rocks.

	Acid	Intermediate		Mafic	Ultramafic
Most Common Minerals	Quartz Orthoclase (and Oligoclase), Mica, Hornblende, Augite	Little or no Quartz: Orthoclase hornblende, Augite, Biotite	Little or no Quartz: Plagioclase Hornblende, Augite, Biotite	No Quartz Plagioclase Augite, Olivine	No Felspar Augite, Hornblende, Olivine
Plutonic or Abyssal type	Granite	Syenite	Diorite	Gabbro	Peridotite
Intrusive or Hypabyssal type	Quartz-porphyry	Orthoclase-porphyry	Porphyrite	Dolerite	Picrite
Lavas or Effusive type	Rhyolite, Obsidian	Trachyte	Andesite	Basalt	Limburgite

Rocks that contain leucite or nepheline, either partly or a wholly replacing felspar, are not included in this table. They are essentially of intermediate or of mafic character. We might in consequence regard them as varieties of syenite, diorite, gabbro, etc., in which feldspathoid minerals occur, and indeed there are many transitions between syenites of ordinary type and nepheline — or leucite — syenite, and between gabbro or dolerite and theralite or essexite. But, as many minerals develop in these "alkali" rocks that are uncommon elsewhere, it is convenient in a purely formal classification like that outlined here to treat the whole assemblage as a distinct series.

Nepheline and Leucite-bearing Rocks			
Most Common Minerals	Alkali Feldspar, Nepheline or Leucite, Augite, Hornblend, Biotite	Soda Lime Feldspar, Nepheline or Leucite, Augite, Hornblende (Olivine)	Nepheline or Leucite, Augite, Hornblende, Olivine
Plutonic type	Nepheline-syenite, Leucite-syenite, Nepheline-porphyry	Essexite and Theralite	Ijolite and Missourite
Effusive type or Lavas	Phonolite, Leucitophyre	Tephrite and Basanite	Nepheline-basalt, Leucite-basalt

This classification is based essentially on the mineralogical constitution of the igneous rocks. Any chemical distinctions between the different groups, though implied, are relegated to a subordinate position. It is admittedly artificial but it has grown up with the growth of the science and is still adopted as the basis on which more minute subdivisions are erected. The subdivisions are by no means of equal value. The syenites, for example, and the peridotites, are far less important than the granites, diorites and gabbros. Moreover, the effusive andesites do not always correspond to the plutonic diorites but partly also to the gabbros. As the different kinds of rock, regarded as aggregates of minerals, pass gradually into one another, transitional types are very common and are often so important as to receive special names. The quartz-syenites and nordmarkites may be interposed between granite and syenite, the tonalites and adamellites between granite and diorite, the monzoaites between syenite and diorite, norites and hyperites between diorite and gabbro, and so on.

Geochemistry of Trace Metals in the Ocean

Trace metals readily form complexes with major ions in the ocean, including hydroxide, carbonate, and chloride and their chemical speciation changes depending on whether the environment is oxidized or reduced. Benjamin (2002) defines complexes of metals with more than one type of ligand, other than water, as mixed-ligand-complexes. In some cases, a ligand contains more than one *donor* atom, forming very strong complexes, also called chelates (the ligand is the chelator). One of the most common chelators is EDTA (ethylenediaminetetraacetic acid), which can replace six molecules of water and form strong bonds with metals that have a plus two charge. With stronger complexation, lower activity of the free metal ion is observed. One consequence of the lower reactivity of complexed metals compared to the same concentration of free metal is that the chelation tends to stabilize metals in the aqueous solution instead of in solids.

Concentrations of the trace metals cadmium, copper, molybdenum, manganese, rhenium, uranium and vanadium in sediments record the redox history of the oceans. Within aquatic environments, cadmium(II) can either be in the form $CdCl^+_{(aq)}$ in oxic waters or $CdS(s)$ in a reduced environment. Thus higher concentrations of Cd in marine sediments may indicate low redox potential conditions in the past. For copper(II), a prevalent form is $CuCl^+(aq)$ within oxic environments and $CuS(s)$ and Cu_2S within reduced environments. The reduced seawater environment leads to two possible oxidation states of copper, Cu(I) and Cu(II). Molybdenum is present as the Mo(VI) oxidation state as $MoO_4^{2-}_{(aq)}$ in oxic environments. Mo(V) and Mo(IV) are present in reduced environments in the forms $MoO_2^+_{(aq)}$ and $MoS_{2(s)}$. Rhenium is present as the Re(VII) oxidation state as ReO_4^- within oxic conditions, but is reduced to Re(IV) which may form ReO_2 or ReS_2. Uranium is in oxidation state VI in $UO_2(CO_3)_3^{4-}(aq)$ and is found in the reduced form $UO_2(s)$. Vanadium is in several forms in oxidation state V(V); HVO_4^{2-} and $H_2VO_4^-$. Its reduced forms can include VO_2^+, $VO(OH)_3^-$, and $V(OH)_3$. These relative dominance of these species depends on pH.

In the water column of the ocean or deep lakes, vertical profiles of dissolved trace metals are characterized as following *conservative–type, nutrient–type,* or *scavenged–type* distributions. Across these three distributions, trace metals have different residence times and are used to varying extents by planktonic microorganisms. Trace metals with conservative-type distributions have high concentrations relative to their biological use. One example of a trace metal with a conservative-type distribution is molybdenum. It has a residence time within the oceans of around 8×10^5 years and is generally present as the molybdate anion (MoO_4^{2-}). Molybdenum interacts weakly with particles and displays an almost uniform vertical profile in the ocean. Relative to the abundance of molybdenum in the ocean, the amount required as a metal cofactor for enzymes in marine phytoplankton is negligible.

Trace metals with nutrient-type distributions are strongly associated with the internal cycles of particulate organic matter, especially the assimilation by plankton. The lowest dissolved concentrations of these metals are at the surface of the ocean, where they are assimilated by plankton. As dissolution and decomposition occur at greater depths, concentrations of these trace metals increase. Residence times of these metals, such as zinc, are several thousand to one hundred thousand years. Finally, an example of a scavenged-type trace metal is aluminium, which has strong interactions with particles as well as a short residence time in the ocean. The residence times of scavenged-type trace metals are around 100 to 1000 years. The concentrations of these metals are highest around bottom sediments, hydrothermal vents, and rivers. For aluminium, atmospheric dust provides the greatest source of external inputs into the ocean.

Iron and copper show hybrid distributions in the ocean. They are influenced by recycling and intense scavenging. Iron is a limiting nutrient in vast areas of the oceans, and is found in high abundance along with manganese near hydrothermal vents. Here, many iron precipitates are found, mostly in the forms of iron sulfides and oxidized iron oxyhydroxide compounds. Concentrations of iron near hydrothermal vents can be up to one million times the concentrations found in the open ocean.

Using electrochemical techniques, it is possible to show that bioactive trace metals (zinc, cobalt, cadmium, iron and copper) are bound by organic ligands in surface seawater. These ligand complexes serve to lower the bioavailability of trace metals within the ocean. For example, copper, which may be toxic to open ocean phytoplankton and bacteria, can form organic complexes. The formation of these complexes reduces the concentrations of bioavailable inorganic complexes of copper that could be toxic to sea life at high concentrations. Unlike copper, zinc toxicity in marine phytoplankton is low and there is no advantage to increasing the organic binding of Zn^{2+}. In high nutrient-low chlorophyll regions, iron is the limiting nutrient, with the dominant species being strong organic complexes of Fe(III).

Geomorphology

Geomorphology (from Greek: γῆ, *gê*, "earth"; μορφή, *morph* , "form"; and λόγος, *lógos*, "study") is the scientific study of the origin and evolution of topographic and bathymetric features created by physical, chemical or biological processes operating at or near the Earth's surface. Geomorphologists seek to understand why landscapes look the way they do, to understand landform history and dynamics and to predict changes through a combination of field observations, physical experiments and numerical modeling. Geomorphologists work within disciplines such as physical geography, geology, geodesy, engineering geology, archaeology and geotechnical engineering. This broad base of interests contributes to many research styles and interests within the field.

Badlands incised into shale at the foot of the North Caineville Plateau, Utah, within the pass carved by the Fremont River and known as the Blue Gate. GK Gilbert studied the landscapes of this area in great detail, forming the observational foundation for many of his studies on geomorphology.

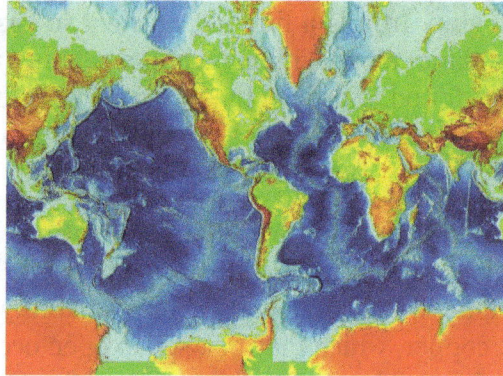

Surface of the Earth, showing higher elevations in red.

Overview

Earth's surface is modified by a combination of surface processes that sculpt landscapes, and geologic processes that cause tectonic uplift and subsidence, and shape the coastal geography. Surface processes comprise the action of water, wind, ice, fire, and living things on the surface of the Earth, along with chemical reactions that form soils and alter material properties, the stability and rate of change of topography under the force of gravity, and other factors, such as (in the very recent past) human alteration of the landscape. Many of these factors are strongly mediated by climate. Geologic processes include the uplift of mountain ranges, the growth of volcanoes, isostatic changes in land surface elevation (sometimes in response to surface processes), and the formation of deep sedimentary basins where the surface of the Earth drops and is filled with material eroded from other parts of the landscape. The Earth's surface and its topography therefore are an intersection of climatic, hydrologic, and biologic action with geologic processes, or alternatively stated, the intersection of the Earth's lithosphere with its hydrosphere, atmosphere, and biosphere.

Waves and water chemistry lead to structural failure in exposed rocks

The broad-scale topographies of the Earth illustrate this intersection of surface and subsurface action. Mountain belts are uplifted due to geologic processes. Denudation of these high uplifted regions produces sediment that is transported and deposited else-where within the landscape or off the coast. On progressively smaller scales, similar ideas apply, where individual landforms evolve in response to the balance of additive processes (uplift and deposition) and subtractive processes (subsidence and erosion). Often, these processes directly affect each other: ice sheets, water, and sediment are all loads that change topography through flexural isostasy. Topography can modify the local climate, for example through orographic precipitation, which in turn modifies the topography by changing the hydrologic regime in which it evolves. Many geomorphologists are particularly interested in the potential for feedbacks between climate and tectonics, mediated by geomorphic processes.

In addition to these broad-scale questions, geomorphologists address issues that are more specific and/or more local. Glacial geomorphologists investigate glacial deposits such as moraines, eskers, and proglacial lakes, as well as glacial erosional features, to build chronologies of both small glaciers and large ice sheets and understand their motions and effects upon the landscape. Fluvial geomorphologists focus on rivers, how they transport sediment, migrate across the landscape, cut into bedrock, respond to environmental and tectonic changes, and interact with humans. Soils geomorphologists investigate soil profiles and chemistry to learn about the history of a particular landscape and understand how climate, biota, and rock interact. Other geomorphologists study how hillslopes form and change. Still others investigate the relationships between ecology and geomorphology. Because geomorphology is defined to comprise everything related to the surface of the Earth and its modification, it is a broad field with many facets.

Geomorphologists use a wide range of techniques in their work. These may include field-work and field data collection, the interpretation of remotely sensed data, geochemical analyses, and the numerical modelling of the physics of landscapes. Geomorphologists may rely on geochronology, using dating methods to measure the rate of changes to the surface. Terrain measurement techniques are vital to quantitatively describe the form of the Earth's surface, and include differential GPS, remotely sensed digital terrain models and laser scanning, to quantify, study, and to generate illustrations and maps.

Practical applications of geomorphology include hazard assessment (such as landslide prediction and mitigation), river control and stream restoration, and coastal protection. Planetary geomorphology studies landforms on other terrestrial planets such as Mars. Indications of effects of wind, fluvial, glacial, mass wasting, meteor impact, tectonics and volcanic processes are studied. This effort not only helps better understand the geologic and atmospheric history of those planets but also extends geomorphological study of the Earth. Planetary geomorphologists often use Earth analogues to aid in their study of surfaces of other planets.

History

Other than some notable exceptions in antiquity, geomorphology is a relatively young science, growing along with interest in other aspects of the earth sciences in the mid-19th century. This section provides a very brief outline of some of the major figures and events in its development.

"Cono de Arita" at the dry lake Salar de Arizaro on the Atacama Plateau, in northwestern Argentina. The cone itself is a volcanic edifice, representing complex interaction of intrusive igneous rocks with the surrounding salt.

Lake "Veľké Hincovo pleso" in High Tatras, Slovakia. The lake occupies an "overdeepening" carved by flowing ice that once occupied this glacial valley.

Ancient Geomorphology

The first theory of geomorphology was arguably devised by the polymath Chinese scientist and statesman Shen Kuo (1031–1095 AD). This was based on his observation of marine fossil shells in a geological stratum of a mountain hundreds of miles from the Pacific Ocean. Noticing bivalve shells running in a horizontal span along the cut section of a cliffside, he theorized that the cliff was once the pre-historic location of a seashore that had shifted hundreds of miles over the centuries. He inferred that the land was reshaped and formed by soil erosion of the mountains and by deposition of silt, after observing strange natural erosions of the Taihang Mountains and the Yandang Mountain near Wenzhou. Furthermore, he promoted the theory of gradual climate change over centuries of time once ancient petrified bamboos were found to be preserved underground in the dry, northern climate zone of *Yanzhou*, which is now modern day Yan'an, Shaanxi province.

Early Modern Geomorphology

The term geomorphology seems to have been first used by Laumann in an 1858 work written in German. Keith Tinkler has suggested that the word came into general use in English, German and French after John Wesley Powell and W. J. McGee used it during the International Geological Conference of 1891. John Edward Marr in his The Scientific Study of Scenery considered his book as, 'an Introductory Treatise on Geomorphology, a subject which has sprung from the union of Geology and Geography'.

An early popular geomorphic model was the *geographical cycle* or *cycle of erosion* model of broad-scale landscape evolution developed by William Morris Davis between 1884 and 1899. It was an elaboration of the uniformitarianism theory that had first been proposed by James Hutton (1726–1797). With regard to valley forms, for example, uniformitarianism posited a sequence in which a river runs through a flat terrain, gradually carving an increasingly deep valley, until the side valleys eventually erode, flattening the terrain again, though at a lower elevation. It was thought that tectonic uplift could then start the cycle over. In the decades following Davis's development of this idea, many of those studying geomorphology sought to fit their findings into this framework, known today as "Davisian". Davis's ideas are of historical importance, but have been largely superseded today, mainly due to their lack of predictive power and qualitative nature.

In the 1920s, Walther Penck developed an alternative model to Davis's. Penck thought that landform evolution was better described as an alternation between ongoing processes of uplift and denudation, as opposed to Davis's model of a single uplift followed by decay. He also emphasised that in many landscapes slope evolution occurs by backwearing of rocks, not by Davisian-style surface lowering, and his science tended to emphasise surface process over understanding in detail the surface history of a given locality. Penck was German, and during his lifetime his ideas were at times rejected vigorously by the English-speaking geomorphology community. His early death, Davis' dislike for his work, and his at-times-confusing writing style likely all contributed to this rejection.

Both Davis and Penck were trying to place the study of the evolution of the Earth's surface on a more generalized, globally relevant footing than it had been previously. In the early 19th century, authors – especially in Europe – had tended to attribute the form of landscapes to local climate, and in particular to the specific effects of glaciation and periglacial processes. In contrast, both Davis and Penck were seeking to emphasize the importance of evolution of landscapes through time and the generality of the Earth's surface processes across different landscapes under different conditions.

During the early 1900s, the study of regional-scale geomorphology was termed "physiography". Physiography later was considered to be a contraction of "*physic*al" and "ge*ography*", and therefore synonymous with physical geography, and the concept became embroiled in controversy surrounding the appropriate concerns of that discipline. Some geomorphologists held to a geological basis for physiography and emphasized a concept of physiographic regions while a conflicting trend among geographers was to equate physiography with "pure morphology," separated from its geological heritage. In the period following World War II, the emergence of process, climatic, and quantitative studies led to a preference by many earth scientists for the term "geomorphology" in order to suggest an analytical approach to landscapes rather than a descriptive one.

Quantitative Geomorphology

Part of the Great Escarpment in the Drakensberg, southern Africa. This landscape, with its high altitude plateau being incised into by the steep slopes of the escarpment, was cited by Davis as a classic example of his cycle of erosion.

Geomorphology was started to be put on a solid quantitative footing in the middle of the 20th century. Following the early work of Grove Karl Gilbert around the turn of the 20th century, a group of natural scientists, geologists and hydraulic engineers including Ralph Alger Bagnold, Hans-Albert Einstein, Frank Ahnert, John Hack, Luna Leopold, A. Shields, Thomas Maddock, Arthur Strahler, Stanley Schumm, and Ronald

Shreve began to research the form of landscape elements such as rivers and hillslopes by taking systematic, direct, quantitative measurements of aspects of them and investigating the scaling of these measurements. These methods began to allow prediction of the past and future behavior of landscapes from present observations, and were later to develop into the modern trend of a highly quantitative approach to geomorphic problems. Quantitative geomorphology can involve fluid dynamics and solid mechanics, geomorphometry, laboratory studies, field measurements, theoretical work, and full landscape evolution modeling. These approaches are used to understand weathering and the formation of soils, sediment transport, landscape change, and the interactions between climate, tectonics, erosion, and deposition.

Contemporary Geomorphology

Today, the field of geomorphology encompasses a very wide range of different approaches and interests. Modern researchers aim to draw out quantitative "laws" that govern Earth surface processes, but equally, recognize the uniqueness of each landscape and environment in which these processes operate. Particularly important realizations in contemporary geomorphology include:

1) that not all landscapes can be considered as either "stable" or "perturbed", where this perturbed state is a temporary displacement away from some ideal target form. Instead, dynamic changes of the landscape are now seen as an essential part of their nature.

2) that many geomorphic systems are best understood in terms of the stochasticity of the processes occurring in them, that is, the probability distributions of event magnitudes and return times. This in turn has indicated the importance of chaotic determinism to landscapes, and that landscape properties are best considered statistically. The same processes in the same landscapes do not always lead to the same end results.

Processes

Gorge cut by the Indus river into bedrock, Nanga Parbat region, Pakistan. This is the deepest river canyon in the world. Nanga Parbat itself, the world's 9th highest mountain, is seen in the background.

Geomorphically relevant processes generally fall into (1) the production of regolith by weathering and erosion, (2) the transport of that material, and (3) its eventual deposition. Primary surface processes responsible for most topographic features include wind, waves, chemical dissolution, mass wasting, groundwater movement, surface water flow, glacial action, tectonism, and volcanism. Other more exotic geomorphic processes might include periglacial (freeze-thaw) processes, salt-mediated action, marine currents activity, seepage of fluids through the seafloor or extraterrestrial impact.

Aeolian Processes

Aeolian processes pertain to the activity of the winds and more specifically, to the winds' ability to shape the surface of the Earth. Winds may erode, transport, and deposit materials, and are effective agents in regions with sparse vegetation and a large supply of fine, unconsolidated sediments. Although water and mass flow tend to mobilize more material than wind in most environments, eolian processes are important in arid environments such as deserts.

Wind-eroded alcove near Moab, Utah

Biological Processes

The interaction of living organisms with landforms, or biogeomorphologic processes, can be of many different forms, and is probably of profound importance for the terrestrial geomorphic system as a whole. Biology can influence very many geomorphic processes, ranging from biogeochemical processes controlling chemical weathering, to the influence of mechanical processes like burrowing and tree throw on soil development, to even controlling global erosion rates through modulation of climate through carbon dioxide balance. Terrestrial landscapes in which the role of biology in mediating surface processes can be definitively excluded are extremely rare, but may hold important information for understanding the geomorphology of other planets, such as Mars.

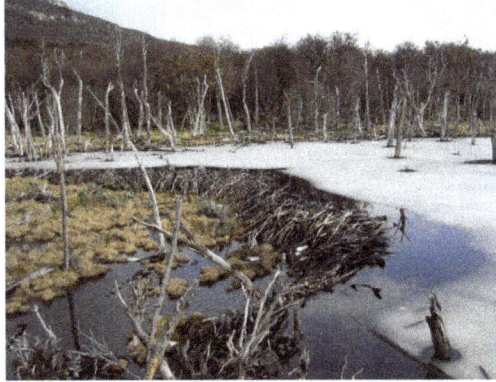

Beaver dams, as this one in Tierra del Fuego, constitute a specific form of zoogeomorphology,
a type of biogeomorphology.

Fluvial Processes

Rivers and streams are not only conduits of water, but also of sediment. The water, as it flows over the channel bed, is able to mobilize sediment and transport it downstream, either as bed load, suspended load or dissolved load. The rate of sediment transport depends on the availability of sediment itself and on the river's discharge. Rivers are also capable of eroding into rock and creating new sediment, both from their own beds and also by coupling to the surrounding hillslopes. In this way, rivers are thought of as setting the base level for large scale landscape evolution in nonglacial environments. Rivers are key links in the connectivity of different landscape elements.

Seif and barchan dunes in the Hellespontus region on the surface of Mars. Dunes are mobile landforms
created by the transport of large volumes of sand by wind.

As rivers flow across the landscape, they generally increase in size, merging with other rivers. The network of rivers thus formed is a drainage system. These systems take on four general patterns, dendritic, radial, rectangular, and trellis. Dendritic happens to be the most common occurring when the underlying strata is stable (without faulting). Drainage systems have four primary components: drainage basin, alluvial valley, delta plain, and receiving basin. Some geomorphic examples of fluvial landforms are alluvial fans, oxbow lakes, and fluvial terraces.

Glacial Processes

Glaciers, while geographically restricted, are effective agents of landscape change. The gradual movement of ice down a valley causes abrasion and plucking of the underlying rock. Abrasion produces fine sediment, termed glacial flour. The debris transported by the glacier, when the glacier recedes, is termed a moraine. Glacial erosion is responsible for U-shaped valleys, as opposed to the V-shaped valleys of fluvial origin.

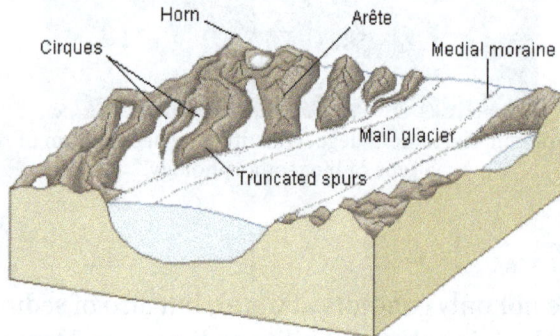

Features of a glacial landscape

The way glacial processes interact with other landscape elements, particularly hillslope and fluvial processes, is an important aspect of Plio-Pleistocene landscape evolution and its sedimentary record in many high mountain environments. Environments that have been relatively recently glaciated but are no longer may still show elevated landscape change rates compared to those that have never been glaciated. Nonglacial geomorphic processes which nevertheless have been conditioned by past glaciation are termed paraglacial processes. This concept contrasts with periglacial processes, which are directly driven by formation or melting of ice or frost.

Hillslope Processes

Soil, regolith, and rock move downslope under the force of gravity via creep, slides, flows, topples, and falls. Such mass wasting occurs on both terrestrial and submarine slopes, and has been observed on Earth, Mars, Venus, Titan and Iapetus.

Talus cones on the north shore of Isfjorden, Svalbard, Norway. Talus cones are accumulations of coarse hillslope debris at the foot of the slopes producing the material.

The Ferguson Slide is an active landslide in the Merced River canyon on California State Highway 140, a primary access road to Yosemite National Park.

Ongoing hillslope processes can change the topology of the hillslope surface, which in turn can change the rates of those processes. Hillslopes that steepen up to certain critical thresholds are capable of shedding extremely large volumes of material very quickly, making hillslope processes an extremely important element of landscapes in tectonically active areas.

On the Earth, biological processes such as burrowing or tree throw may play important roles in setting the rates of some hillslope processes.

Igneous Processes

Both volcanic (eruptive) and plutonic (intrusive) igneous processes can have important impacts on geomorphology. The action of volcanoes tends to rejuvenate landscapes, covering the old land surface with lava and tephra, releasing pyroclastic material and forcing rivers through new paths. The cones built by eruptions also build substantial new topography, which can be acted upon by other surface processes. Plutonic rocks intruding then solidifying at depth can cause both uplift or subsidence of the surface, depending on whether the new material is denser or less dense than the rock it displaces.

Tectonic Processes

Tectonic effects on geomorphology can range from scales of millions of years to minutes or less. The effects of tectonics on landscape are heavily dependent on the nature of the underlying bedrock fabric that more less controls what kind of local morphology tectonics can shape. Earthquakes can, in terms of minutes, submerge large areas of land creating new wetlands. Isostatic rebound can account for significant changes over hundreds to thousands of years, and allows erosion of a mountain belt to promote further erosion as mass is removed from the chain and the belt uplifts. Long-term plate tectonic dynamics give rise to orogenic belts, large mountain chains with typical lifetimes of many tens of millions of years, which form focal points for high rates of fluvial and hillslope processes and thus long-term sediment production.

Features of deeper mantle dynamics such as plumes and delamination of the lower lithosphere have also been hypothesised to play important roles in the long term (> million year), large scale (thousands of km) evolution of the Earth's topography. Both can promote surface uplift through isostasy as hotter, less dense, mantle rocks displace cooler, denser, mantle rocks at depth in the Earth.

Marine Processes

Marine processes are those associated with the action of waves, marine currents and seepage of fluids through the seafloor.

Scales in Geomorphology

Different geomorphological processes dominate at different spatial and temporal scales. Moreover, scales on which processes occur may determine the reactivity or otherwise of landscapes to changes in driving forces such as climate or tectonics. These ideas are key to the study of geomorphology today.

To help categorize landscape scales some geomorphologists might use the following taxonomy:

- 1st – Continent, ocean basin, climatic zone ($\sim$10,000,000 km^2)
- 2nd – Shield, e.g. Baltic Shield, or mountain range ($\sim$1,000,000 km^2)
- 3rd – Isolated sea, Sahel ($\sim$100,000 km^2)
- 4th – Massif, e.g. Massif Central or Group of related landforms, e.g., Weald ($\sim$10,000 km^2)
- 5th – River valley, Cotswolds ($\sim$1,000 km^2)
- 6th – Individual mountain or volcano, small valleys ($\sim$100 km^2)
- 7th – Hillslopes, stream channels, estuary ($\sim$10 km^2)
- 8th – gully, barchannel ($\sim$1 km^2)
- 9th – Meter-sized features

Overlap with Other Fields

There is a considerable overlap between geomorphology and other fields. Deposition of material is extremely important in sedimentology. Weathering is the chemical and physical disruption of earth materials in place on exposure to atmospheric or near surface agents, and is typically studied by soil scientists and environmental chemists, but is an essential component of geomorphology because it is what provides the material that can be moved in the first place. Civil and environmental engineers are concerned with erosion and sediment transport, especially related to canals, slope stability (and natural hazards), water quality, coastal environmental management, transport of contaminants, and stream restoration. Glaciers can cause extensive erosion and deposition in a short period of time, making them extremely important entities in the high latitudes and meaning that they set the conditions in the headwaters of mountain-born streams; glaciology therefore is important in geomorphology.

Isotope Geochemistry

Isotope geochemistry is an aspect of geology based upon study of the natural variations in the relative abundances of isotopes of various elements. Variations in isotopic abundance are measured by isotope ratio mass spectrometry, and can reveal information about the ages and origins of rock, air or water bodies, or processes of mixing between them.

Stable isotope geochemistry is largely concerned with isotopic variations arising from mass-dependent isotope fractionation, whereas radiogenic isotope geochemistry is concerned with the products of natural radioactivity.

Stable Isotope Geochemistry

For most stable isotopes, the magnitude of fractionation from kinetic and equilibrium fractionation is very small; for this reason, enrichments are typically reported in "per mil" (‰, parts per thousand). These enrichments (δ) represent the ratio of heavy isotope to light isotope in the sample over the ratio of a standard.

Hydrogen

Hydrogen isotope biogeochemistry

Carbon

Carbon has two stable isotopes, ^{12}C and ^{13}C, and one radioactive isotope, ^{14}C.

The stable carbon isotope ratio, $\delta^{13}C$, is measured against Vienna Pee Dee Belemnite (VPDB). The stable carbon isotopes are fractionated primarily by photosynthesis (Faure, 2004). The $^{13}C/^{12}C$ ratio is also an indicator of paleoclimate: a change in the ratio in the remains of plants indicates a change in the amount of photosynthetic activity, and thus in how favorable the environment was for the plants. During photosynthesis, organisms using the C_3 pathway show different enrichments compared to those using the C_4 pathway, allowing scientists not only to distinguish organic matter from abiotic carbon, but also what type of photosynthetic pathway the organic matter was using. Occasional spikes in the global $^{13}C/^{12}C$ ratio have also been useful as stratigraphic markers for chemostratigraphy, especially during the Paleozoic.

The ^{14}C ratio has been used to track ocean circulation, among other things.

Nitrogen

Nitrogen has two stable isotopes, ^{14}N, and ^{15}N. The ratio between these is measured relative to nitrogen in ambient air. Nitrogen ratios are frequently linked to agricultural activities. Nitrogen isotope data has also been used to measure the amount of exchange of air between the stratosphere and troposphere using data from the greenhouse gas N_2O.

Oxygen

Oxygen has three stable isotopes, ^{16}O, ^{17}O, and ^{18}O. Oxygen ratios are measured relative to Vienna Standard Mean Ocean Water (VSMOW) or Vienna Pee Dee Belemnite (VPDB). Variations in oxygen isotope ratios are used to track both water movement, paleoclimate, and atmospheric gases such as ozone and carbon dioxide. Typically, the VPDB oxygen reference is used for paleoclimate, while VSMOW is used for most other applications. Oxygen isotopes appear in anomalous ratios in atmospheric ozone, resulting from mass-independent fractionation. Isotope ratios in fossilized foraminifera have been used to deduce the temperature of ancient seas.

Sulfur

Sulfur has four stable isotopes, with the following abundances: ^{32}S (0.9502), ^{33}S (0.0075), ^{34}S (0.0421) and ^{36}S (0.0002). These abundances are compared to those found in Cañon Diablo troilite. Variations in sulfur isotope ratios are used to study the origin of sulfur in an orebody and the temperature of formation of sulfur–bearing minerals.

Radiogenic Isotope Geochemistry

Radiogenic isotopes provide powerful tracers for studying the ages and origins of Earth systems. They are particularly useful to understand mixing processes between different components, because (heavy) radiogenic isotope ratios are not usually fractionated by chemical processes.

Radiogenic isotope tracers are most powerful when used together with other tracers: The more tracers used, the more control on mixing processes. An example of this application is to the evolution of the Earth's crust and Earth's mantle through geological time.

Lead-lead Isotope Geochemistry

Lead has four stable isotopes - ^{204}Pb, ^{206}Pb, ^{207}Pb, ^{208}Pb and one common radioactive isotope ^{202}Pb with a half-life of ~53,000 years.

Lead is created in the Earth via decay of transuranic elements, primarily uranium and thorium.

Lead isotope geochemistry is useful for providing isotopic dates on a variety of materials. Because the lead isotopes are created by decay of different transuranic elements, the ratios of the four lead isotopes to one another can be very useful in tracking the source of melts in igneous rocks, the source of sediments and even the origin of people via isotopic fingerprinting of their teeth, skin and bones.

It has been used to date ice cores from the Arctic shelf, and provides information on the source of atmospheric lead pollution.

Lead-lead isotopes has been successfully used in forensic science to fingerprint bullets, because each batch of ammunition has its own peculiar $^{204}Pb/^{206}Pb$ vs $^{207}Pb/^{208}Pb$ ratio.

Samarium-neodymium

Samarium-neodymium is an isotope system which can be utilised to provide a date as well as isotopic fingerprints of geological materials, and various other materials including archaeological finds (pots, ceramics).

^{147}Sm decays to produce ^{143}Nd with a half life of 1.06×10^{11} years.

Dating is achieved usually by trying to produce an isochron of several minerals within a rock specimen. The initial $^{143}Nd/^{144}Nd$ ratio is determined.

This initial ratio is modelled relative to CHUR - the Chondritic Uniform Reservoir - which is an approximation of the chondritic material which formed the solar system. CHUR was determined by analysing chondrite and achondrite meteorites.

The difference in the ratio of the sample relative to CHUR can give information on a model age of extraction from the mantle (for which an assumed evolution has been calculated relative to CHUR) and to whether this was extracted from a granitic source (depleted in radiogenic Nd), the mantle, or an enriched source.

Rhenium-osmium

Rhenium and osmium are siderophile elements which are present at very low abundances in the crust. Rhenium undergoes radioactive decay to produce osmium. The ratio of non-radiogenic osmium to radiogenic osmium throughout time varies.

Rhenium prefers to enter sulfides more readily than osmium. Hence, during melting of the mantle, rhenium is stripped out, and prevents the osmium-osmium ratio from changing appreciably. This *locks in* an initial osmium ratio of the sample at the time of the melting event. Osmium-osmium initial ratios are used to determine the source characteristic and age of mantle melting events.

Noble Gas Isotopes

Natural isotopic variations amongst the noble gases result from both radiogenic and nucleogenic production processes. Because of their unique properties, it is useful to distinguish them from the conventional radiogenic isotope systems described above.

Helium-3

Helium-3 was trapped in the planet when it formed. Some ^{3}He is being added by meteoric dust, primarily collecting on the bottom of oceans (although due to subduction, all oceanic tectonic plates are younger than continental plates). However, ^{3}He will be degassed from oceanic sediment during subduction, so cosmogenic ^{3}He is not affecting the concentration or noble gas ratios of the mantle.

Helium-3 is created by cosmic ray bombardment, and by lithium spallation reactions which generally occur in the crust. Lithium spallation is the process by which a high-energy neutron bombards a lithium atom, creating a ^{3}He and a ^{4}He ion. This requires significant lithium to adversely affect the ^{3}He/^{4}He ratio.

All degassed helium is lost to space eventually, due to the average speed of helium exceeding the escape velocity for the Earth. Thus, it is assumed the helium content and ratios of Earth's atmosphere have remained essentially stable.

It has been observed that ^{3}He is present in volcano emissions and oceanic ridge samples. How ^{3}He is stored in the planet is under investigation, but it is associated with the mantle and is used as a marker of material of deep origin.

Due to similarities in helium and carbon in magma chemistry, outgassing of helium requires the loss of volatile components (water, carbon dioxide) from the mantle, which happens at depths of less than 60 km. However, ^{3}He is transported to the surface primarily trapped in the crystal lattice of minerals within fluid inclusions.

Helium-4 is created by radiogenic production (by decay of uranium/thorium-series elements). The continental crust has become enriched with those elements relative to the mantle and thus more He^4 is produced in the crust than in the mantle.

The ratio (R) of ^{3}He to ^{4}He is often used to represent ^{3}He content. R usually is given as a multiple of the present atmospheric ratio (Ra).

Common values for R/Ra:

- Old continental crust: less than 1

- mid-ocean ridge basalt (MORB): 7 to 9

- Spreading ridge rocks: 9.1 plus or minus 3.6

- Hotspot rocks: 5 to 42

- Ocean and terrestrial water: 1

- Sedimentary formation water: less than 1

- Thermal spring water: 3 to 11

^{3}He/^{4}He isotope chemistry is being used to date groundwaters, estimate groundwater flow rates, track water pollution, and provide insights into hydrothermal processes, igneous geology and ore genesis.

- (U-Th)/He dating of apatite as a thermal history tool

- USGS: Helium Discharge at Mammoth Mountain Fumarole (MMF)

Uranium-series Isotopes

U-series isotopes are unique amongst radiogenic isotopes because, being in the U-series decay chains, they are both radiogenic and radioactive. Because their abundances are normally quoted as activity ratios rather than atomic ratios, they are best considered separately from the other radiogenic isotope systems.

Protactinium/Thorium - ^{231}Pa / ^{230}Th

Uranium is well mixed in the ocean, and its decay produces ^{231}Pa and ^{230}Th at a constant activity ratio (0.093). The decay products are rapidly removed by adsorption on settling particles, but not at equal rates. ^{231}Pa has a residence equivalent to the residence time of deep water in the Atlantic basin (around 1000 yrs) but ^{230}Th is removed more rapidly (centuries). Thermohaline circulation effectively exports ^{231}Pa from the Atlantic into the Southern Ocean, while most of the ^{230}Th remains in Atlantic sediments. As a result, there is a relationship between ^{231}Pa/^{230}Th in Atlantic sediments and the rate of overturning: faster overturning produces lower sediment ^{231}Pa/^{230}Th ratio, while slower overturning increases this ratio. The combination of δ^{13}C and ^{231}Pa/^{230}Th can therefore provide a more complete insight into past circulation changes.

Anthropogenic Isotopes

Tritium/Helium-3

Tritium was released to the atmosphere during atmospheric testing of nuclear bombs. Radioactive decay of tritium produces the noble gas helium-3. Comparing the ratio of tritium to helium-3 (^{3}H/^{3}He) allows estimation of the age of recent ground waters.

- USGS Tritium/Helium-3 Dating

- Hydrologic Isotope Tracers - Helium

Biogeochemistry

Biogeochemistry is the scientific discipline that involves the study of the chemical, physical, geological, and biological processes and reactions that govern the composition of the natural environment (including the biosphere, the cryosphere, the hydrosphere, the pedosphere, the atmosphere, and the lithosphere). In particular, biogeochemistry is the study of the cycles of chemical elements, such as carbon and nitrogen, and their interactions with and incorporation into living things transported through earth scale biological systems in space through time. The field focuses on chemical cycles which are either driven by or influence biological activity. Particular emphasis is placed on the study of carbon, nitrogen, sulfur, and phosphorus cycles. Biogeochemistry is a systems science closely related to systems ecology.

Vladimir Vernadsky
- Russian geochemist.

The founder of biogeochemistry was Ukrainian scientist Vladimir Vernadsky whose 1926 book *The Biosphere*, in the tradition of Mendeleev, formulated a physics of the earth as a living whole. Vernadsky distinguished three spheres, where a sphere was a concept similar to the concept of a phase-space. He observed that each sphere had its own laws of evolution, and that the higher spheres modified and dominated the lower:

1. Abiotic sphere - all the non-living energy and material processes

2. Biosphere - the life processes that live within the abiotic sphere

3. Nöesis or Nösphere - the sphere of the cognitive process of man

Human activities (e.g., agriculture and industry) modify the Biosphere and Abiotic sphere. In the contemporary environment, the amount of influence humans have on the other two spheres is comparable to a geological force.

Early Development

The American limnologist and geochemist G. Evelyn Hutchinson is credited with outlining the broad scope and principles of this new field. More recently, the basic elements of the discipline of biogeochemistry were restated and popularized by the British scientist and writer, James Lovelock, under the label of the *Gaia Hypothesis*. Lovelock emphasizes a concept that life processes regulate the Earth through feedback mechanisms to keep it habitable.

Research

There are biogeochemistry research groups in many universities around the world. Since this is a highly inter-disciplinary field, these are situated within a wide range of host disciplines including: atmospheric sciences, biology, ecology, geomicrobiology, environmental chemistry, geology, oceanography and soil science. These are often bracketed into larger disciplines such as earth science and environmental science.

Many researchers investigate the biogeochemical cycles of chemical elements such as carbon, oxygen, nitrogen, phosphorus and sulfur, as well as their stable isotopes. The cycles of trace elements such as the trace metals and the radionuclides are also studied. This research has obvious applications in the exploration for ore deposits and oil, and in remediation of environmental pollution.

Some important research fields for biogeochemistry include:

- modelling of natural systems

- soil and water acidification recovery processes

- eutrophication of surface waters

- carbon sequestration

- soil remediation

- global change

- climate change

- biogeochemical prospecting for ore deposits

References

- Terzaghi, K., Peck, R.B. and Mesri, G. (1996), Soil Mechanics in Engineering Practice 3rd Ed., John Wiley & Sons, Inc. ISBN 0-471-08658-4

- Soil Mechanics, Lambe,T.William and Whitman,Robert V., Massachusetts Institute of Technology, John Wiley & Sons., 1969. ISBN 0-471-51192-7

- Coduto, Donald; et al. (2011). Geotechnical Engineering Principles and Practices. New Jersey: Pearson Higher Education. ISBN 9780132368681.

- RAJU, V. R. (2010). Ground Improvement Technologies and Case Histories. Singapore: Research Publishing Services. p. 809. ISBN 978-981-08-3124-0.

- Davies, Geoffrey F. (2001). Dynamic Earth: Plates, Plumes and Mantle Convection. Cambridge University Press. ISBN 0-521-59067-1.

- Eratosthenes (2010). Eratosthenes' "Geography". Fragments collected and translated, with commentary and additional material by Duane W. Roller. Princeton University Press. ISBN 978-0-691-14267-8.

- Fowler, C.M.R. (2005). The Solid Earth: An Introduction to Global Geophysics (2 ed.). Cambridge University Press. ISBN 0-521-89307-0.

- Poirier, Jean-Paul (2000). Introduction to the Physics of the Earth's Interior. Cambridge Topics in Mineral Physics & Chemistry. Cambridge University Press. ISBN 0-521-66313-X.

- Sheriff, Robert E. (1991). "Geophysics". Encyclopedic Dictionary of Exploration Geophysics (3rd ed.). Society of Exploration. ISBN 978-1-56080-018-7.

- Stein, Seth; Wysession, Michael (2003). An introduction to seismology, earthquakes, and earth structure. Wiley-Blackwell. ISBN 0-86542-078-5.

- Telford, William Murray; Geldart, L. P.; Sheriff, Robert E. (1990). Applied geophysics. Cambridge University Press. ISBN 978-0-521-33938-4.

Various Geological Processes

Geology has a number of processes; some of these are shear, fold, saltation, metasomatism, denudation and spheroidal weathering. Geological fold occurs when the sedimentary strata becomes bent generally because of a permanent distortion. Saltation is another process related to geology that takes place with the help of wind or water, which help in the transportation of particles.

Fold (Geology)

A geological fold occurs when one or a stack of originally flat and planar surfaces, such as sedimentary strata, are bent or curved as a result of permanent deformation. Synsedimentary folds are those due to slumping of sedimentary material before it is lithified. Folds in rocks vary in size from microscopic crinkles to mountain-sized folds. They occur singly as isolated folds and in extensive fold trains of different sizes, on a variety of scales.

Folds in paleoproterozoic marble in Nunavut, Canada (with hammer for scale).

Folds form under varied conditions of stress, hydrostatic pressure, pore pressure, and temperature gradient, as evidenced by their presence in soft sediments, the full spectrum of metamorphic rocks, and even as primary flow structures in some igneous rocks. A set of folds distributed on a regional scale constitutes a fold belt, a common feature of orogenic zones. Folds are commonly formed by shortening of existing layers, but may also be formed as a result of displacement on a non-planar fault (*fault bend fold*), at the tip of a propagating fault (*fault propagation fold*), by differential compaction or due to the effects of a high-level igneous intrusion e.g. above a laccolith.

Folds in alternating layers of limestone and chert in Crete, Greece.

Describing Folds

Folds are classified by their size, fold shape, tightness, and dip of the axial plane.

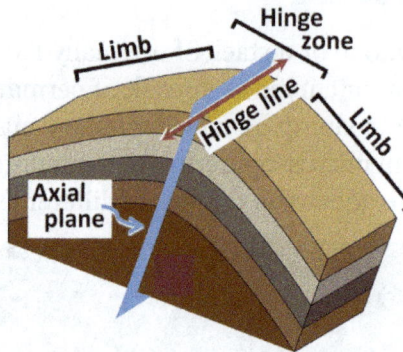

Fold terminology. For more general fold shapes, a hinge *curve* replaces the hinge line, and a non-planar axial *surface* replaces the axial plane.

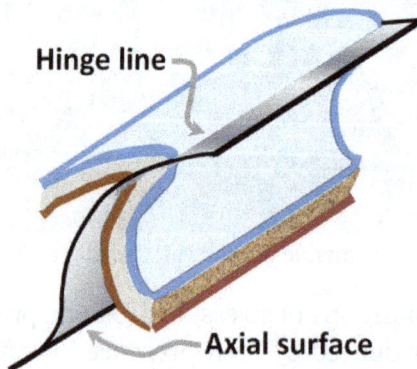

Cylindrical fold with axial surface not a plane.

Fold Terminology in Two Dimensions

A fold surface seen in profile can be divided into *hinge* and *limb* portions. The limbs are the flanks of the fold and the hinge is where the flanks join together. The hinge point is the point of minimum radius of curvature (maximum curvature) for a fold. The crest of the fold is the highest point of the fold surface, and the trough is the lowest point. The inflection point of a fold is the point on a limb at which the concavity reverses; on regular folds, this is the midpoint of the limb.

Fold Terminology in Three Dimensions

The hinge points along an entire folded surface form a hinge line, which can be either a *crest line* or a *trough line*. The trend and plunge of a linear hinge line gives you information about the orientation of the fold. To more completely describe the orientation of a fold, one must describe the *axial surface*. The axial surface is the surface defined by connecting all the hinge lines of stacked folding surfaces. If the axial surface is a planar surface then it is called the axial plane and can be described by the strike and dip of the plane. An *axial trace* is the line of intersection of the axial surface with any other surface (ground, side of mountain, geological cross-section).

Finally, folds can have, but don't necessarily have a *fold axis*. A fold axis, "is the closest approximation to a straight line that when moved parallel to itself, generates the form of the fold." (Davis and Reynolds, 1996 after Donath and Parker, 1964; Ramsay 1967). A fold that can be generated by a fold axis is called a *cylindrical fold*. This term has been broadened to include near-cylindrical folds. Often, the fold axis is the same as the hinge line.

Fold Shape

A fold can be shaped as a chevron, with planar limbs meeting at an angular axis, as cuspate with curved limbs, as circular with a curved axis, or as elliptical with unequal wavelength.

Fold Tightness

Fold tightness is defined by the size of the angle between the fold's limbs (as measured tangential to the folded surface at the inflection line of each limb), called the interlimb angle. Gentle folds have an interlimb angle of between 180° and 120°, open folds range from 120° to 70°, close folds from 70° to 30°, and tight folds from 30° to 0°. *Isoclines*, or *isoclinal folds*, have an interlimb angle of between 10° and zero, with essentially parallel limbs.

Fold Symmetry

Not all folds are equal on both sides of the axis of the fold. Those with limbs of relatively equal length are termed symmetrical, and those with highly unequal limbs are asymmetrical. Asymmetrical folds generally have an axis at an angle to the original unfolded surface they formed on.

Deformation Style Classes

Folds that maintain uniform layer thickness are classed as *concentric* folds. Those that do not are called *similar folds*. Similar folds tend to display thinning of the limbs and thickening of the hinge zone. Concentric folds are caused by warping from active buckling of the layers, whereas similar folds usually form by some form of shear flow where the layers are not mechanically active. Ramsay has proposed a classification scheme for folds that often is used to describe folds in profile based upon curvature of the inner and outer lines of a fold, and the behavior of *dip isogons*. that is, lines connecting points of equal dip:

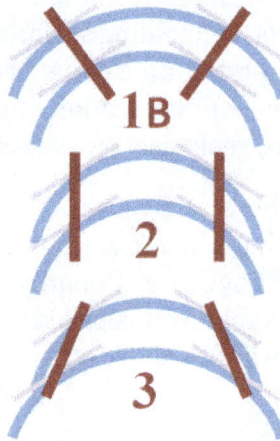

Ramsay classification of folds by convergence of dip isogons (red lines).

Ramsay classification scheme for folds		
Class	**Curvature C**	**Comment**
1	$C_{inner} > C_{outer}$	Dip isogons converge
1A		Orthogonal thickness at hinge narrower than at limbs
1B		Parallel folds
1C		Orthogonal thickness at limbs narrower than at hinge
2	$C_{inner} = C_{outer}$	Dip isogons are parallel: **similar** folds
3	$C_{inner} < C_{outer}$	Dip isogons diverge

Fold Types

Anticline

Monocline at Colorado National Monument

- Anticline: linear, strata normally dip away from axial center, *oldest* strata in center.

- Syncline: linear, strata normally dip toward axial center, *youngest* strata in center.

- Antiform: linear, strata dip away from axial center, age unknown, or inverted.

- Synform: linear, strata dip toward axial centre, age unknown, or inverted.

- Dome: nonlinear, strata dip away from center in all directions, *oldest* strata in center.

- Basin: nonlinear, strata dip toward center in all directions, *youngest* strata in center.

- Monocline: linear, strata dip in one direction between horizontal layers on each side.

- Chevron: angular fold with straight limbs and small hinges

- Recumbent: linear, fold axial plane oriented at low angle resulting in overturned strata in one limb of the fold.

- Slump: typically monoclinal, result of differential compaction or dissolution during sedimentation and lithification.

- Ptygmatic: Folds are chaotic, random and disconnected. Typical of sedimentary slump folding, migmatites and decollement detachment zones.

- Parasitic: short wavelength folds formed within a larger wavelength fold structure - normally associated with differences in bed thickness

- Disharmonic: Folds in adjacent layers with different wavelengths and shapes

(A homocline involves strata dipping in the same direction, though not necessarily any folding.)

Causes of Folding

Folds appear on all scales, in all rock types, at all levels in the crust and arise from a variety of causes.

Layer-parallel Shortening

Box fold in La Herradura Formation, Morro Solar, Peru.

When a sequence of layered rocks is shortened parallel to its layering, this deformation may be accommodated in a number of ways, homogeneous shortening, reverse faulting or folding. The response depends on the thickness of the mechanical layering and the contrast in properties between the layers. If the layering does begin to fold, the fold style is also dependent on these properties. Isolated thick competent layers in a less competent matrix control the folding and typically generate classic rounded buckle folds accommodated by deformation in the matrix. In the case of regular alternations of layers of contrasting properties, such as sandstone-shale sequences, kink-bands, box-folds and chevron folds are normally produced.

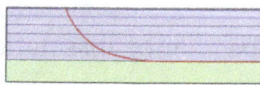

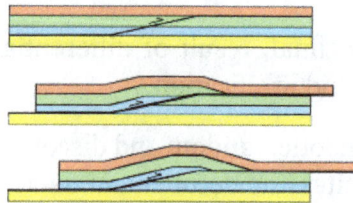

Rollover anticline   Ramp anticline   Fault-propagation fold

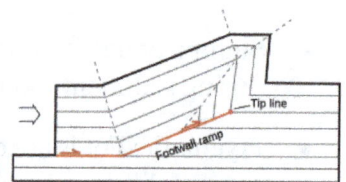

Fault-related Folding

Many folds are directly related to faults, associate with their propagation, displacement and the accommodation of strains between neighbouring faults.

Fault Bend Folding

Fault bend folds are caused by displacement along a non-planar fault. In non-vertical faults, the hanging-wall deforms to accommodate the mismatch across the fault as displacement progresses. Fault bend folds occur in both extensional and thrust faulting. In extension, listric faults form rollover anticlines in their hanging walls. In thrusting, *ramp anticlines* are formed whenever a thrust fault cuts up section from one detachment level to another. Displacement over this higher-angle ramp generates the folding.

Fault Propagation Folding

Fault propagation folds or *tip-line folds* are caused when displacement occurs on an existing fault without further propagation. In both reverse and normal faults this leads to folding of the overlying sequence, often in the form of a monocline.

Detachment Folding

When a thrust fault continues to displace above a planar detachment without further fault propagation, detachment folds may form, typically of box-fold style. These generally occur above a good detachment such as in the Jura Mountains, where the detachment occurs on middle Triassic evaporites.

Folding in Shear Zones

Shear zones that approximate to simple shear typically contain minor asymmetric folds, with the direction of overturning consistent with the overall shear sense. Some of these folds have highly curved hinge lines and are referred to as *sheath folds*. Folds in shear zones can be inherited, formed due to the orientation of pre-shearing layering or formed due to instability within the shear flow.

Dextral sense shear folds in mylonites within a shear zone, Cap de Creus

Folding in Sediments

Recently deposited sediments are normally mechanically weak and prone to remobilisation before they become lithified, leading to folding. To distinguish them from folds of tectonic origin, such structures are called synsedimentary (formed during sedimentation).

Slump folding: When slumps form in poorly consolidated sediments, they commonly undergo folding, particularly at their leading edges, during their emplacement. The asymmetry of the slump folds can be used to determine paleoslope directions in sequences of sedimentary rocks.

Dewatering: Rapid dewatering of sandy sediments, possibly triggered by seismic activity, can cause convolute bedding.

Compaction: Folds can be generated in a younger sequence by differential compaction over older structures such as fault blocks and reefs.

Igneous Intrusion

The emplacement of igneous intrusions tends to deform the surrounding country rock. In the case of high-level intrusions, near the Earth's surface, this deformation is concentrated above the intrusion and often takes the form of folding, as with the upper surface of a laccolith.

Flow Folding

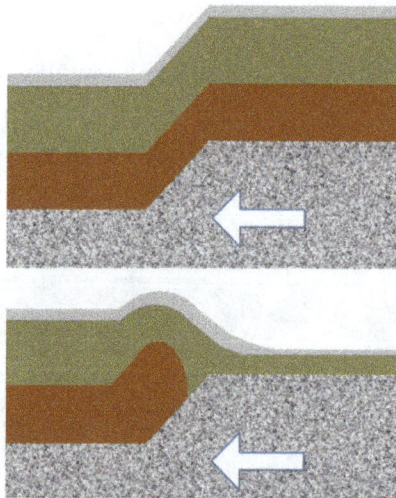

Flow folding: this picture uses artistic license to show the effect of an advancing ramp of rigid rock into compliant layers. Top: low drag by ramp: layers are not altered in thickness; Bottom: high drag: lowest layers tend to crumple.

The compliance of rock layers is referred to as *competence*: a competent layer or bed of rock can withstand an applied load without collapsing and is relatively strong, while an incompetent layer is relatively weak. When rock behaves as a fluid, as in the case of very weak rock such as rock salt, or any rock that is buried deeply enough, it typically shows *flow folding* (also called *passive folding*, because little resistance is offered): the strata appear shifted undistorted, assuming any shape impressed upon them by surrounding more rigid rocks. The strata simply serve as markers of the folding. Such folding is also a feature of many igneous intrusions and glacier ice.

Folding Mechanisms

Folding of rocks must balance the deformation of layers with the conservation of volume in a rock mass. This occurs by several mechanisms.

Example of a large-scale crenulation, an example of chevron-type flexural-slip folds in the Glengarry Basin, W.A.

Flexural Slip

Flexural slip allows folding by creating layer-parallel slip between the layers of the folded strata, which, altogether, result in deformation. A good analogy is bending a phone book, where volume preservation is accommodated by slip between the pages of the book.

The fold formed by the compression of competent rock beds is called "flexure fold".

Buckling

Typically, folding is thought to occur by simple buckling of a planar surface and its confining volume. The volume change is accommodated by *layer parallel shortening* the volume, which grows in *thickness*. Folding under this mechanism is typically of the similar fold style, as thinned limbs are shortened horizontally and thickened hinges do so vertically.

Mass Displacement

If the folding deformation cannot be accommodated by flexural slip or volume-change shortening (buckling), the rocks are generally removed from the path of the stress. This is achieved by pressure dissolution, a form of metamorphic process, in which rocks shorten by dissolving constituents in areas of high strain and redepositing them in areas of lower strain. Folds created in this way include examples in migmatites, and areas with a strong axial planar cleavage.

Mechanics of Folding

Folds in rock are formed in relation to the stress field in which the rocks are located and the rheology, or method of response to stress, of the rock at the time at which the stress is applied.

The rheology of the layers being folded determines characteristic features of the folds that are measured in the field. Rocks that deform more easily form many short-wavelength, high-amplitude folds. Rocks that do not deform as easily form long-wavelength, low-amplitude folds.

Shear (Geology)

Shear is the response of a rock to deformation usually by compressive stress and forms particular textures. Shear can be homogeneous or non-homogeneous, and may be pure shear or simple shear. Study of geological shear is related to the study of structural geology, rock microstructure or rock texture and fault mechanics.

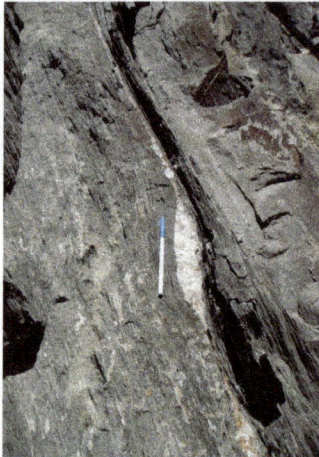

Boudinaged quartz vein(with strain fringe) showing sinistral shear sense, Starlight Pit, Fortnum Gold Mine, Western Australia

The process of shearing occurs within brittle, brittle-ductile, and ductile rocks. Within purely brittle rocks, compressive stress results in fracturing and simple faulting.

Rocks

Rocks typical of shear zones include mylonite, cataclasite, S-tectonite and L-tectonite, pseudotachylite, certain breccias and highly foliated versions of the wall rocks.

Shear Zone

A shear zone is a tabular to sheetlike, planar or curviplanar zone composed of rocks that are more highly strained than rocks adjacent to the zone. Typically this is a type of fault, but it may be difficult to place a distinct fault plane into the shear zone. Shear zones may form zones of much more intense foliation, deformation, and folding. En echelon veins or fractures may be observed within shear zones.

Asymmetric shear in basalt, Labouchere mine, Glengarry Basin, Australia.
Shear asymmetry is sinistral, pen for scale

Many shear zones host ore deposits as they are a focus for hydrothermal flow through orogenic belts. They may often show some form of retrograde metamorphism from a peak metamorphic assemblage and are commonly metasomatised.

Shear zones can be only inches wide, or up to several kilometres wide. Often, due to their structural control and presence at the edges of tectonic blocks, shear zones are mappable units and form important discontinuities to separate terranes. As such, many large and long shear zones are named, identical to fault systems.

When the horizontal displacement of this faulting can be measured in the tens or hundreds of kilometers of length, the fault is referred to as a megashear. Megashears often indicate the edges of ancient tectonic plates.

Mechanisms of Shearing

The mechanisms of shearing depend on the pressure and temperature of the rock and on the rate of shear which the rock is subjected to. The response of the rock to these conditions determines how it accommodates the deformation.

Shear zones which occur in more brittle rheological conditions (cooler, less confining pressure) or at high rates of strain, tend to fail by brittle failure; breaking of minerals, which are ground up into a breccia with a *milled* texture.

Shear zones which occur under brittle-ductile conditions can accommodate much deformation by enacting a series of mechanisms which rely less on fracture of the rock and occur within the minerals and the mineral lattices themselves. Shear zones accommodate compressive stress by movement on foliation planes.

Dextral slickenside of pyrite

Shearing at ductile conditions may occur by fracturing of minerals and growth of subgrain boundaries, as well as by *lattice glide*. This occurs particularly on platy minerals, especially micas.

Mylonites are essentially ductile shear zones.

Microstructures of Shear Zones

During the initiation of shearing, a penetrative planar foliation is first formed within the rock mass. This manifests as realignment of textural features, growth and realignment of micas and growth of new minerals.

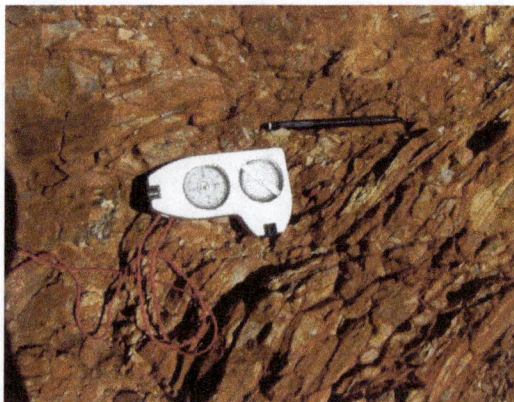

Typical example of dextral shear foliation in an L-S tectonite, with pencil pointing in direction of shear sense. Note the sinusoidal nature of the shear foliation.

The incipient shear foliation typically forms normal to the direction of principal shortening, and is diagnostic of the direction of shortening. In symmetric shortening, objects flatten on this shear foliation much the same way that a round ball of treacle flattens with gravity.

Within asymmetric shear zones, the behavior of an object undergoing shortening is analogous to the ball of treacle being smeared as it flattens, generally into an ellipse. Within shear zones with pronounced displacements a shear foliation may form at a shallow angle to the gross plane of the shear zone. This foliation ideally manifests as a sinusoidal set of foliations formed at a shallow angle to the main shear foliation, and which curve into the main shear foliation. Such rocks are known as L-S tectonites.

If the rock mass begins to undergo large degrees of lateral movement, the strain ellipse lengthens into a cigar shaped volume. At this point shear foliaions begin to break down into a rodding lineation or a stretch lineation. Such rocks are known as L-tectonites.

Stretched pebble conglomerate L-tectonite illustrating a stretch lineation within a shear zone, Glengarry Basin, Australia. Pronounced asymmetric shearing has stretched the conglomerate pebbles into elongate cigar shaped rods.

Ductile Shear Microstructures

Thin section (crossed polars) of Garnet-Mica-Schist showing a rotated porphyroblast of garnet, mica fish and elongated minerals. This specimen was from close to a shear zone in Norway (the Ose thrust), the garnet in the centre (black) is approximately 2mm in diameter

Very distinctive textures form as a consequence of ductile shear. An important group of microstructures observed in ductile shear zones are S-planes, C-planes and C' planes.

- S-planes or *schistosité* planes are generally defined by a planar fabric caused by the alignment of micas or platy minerals. Define the flattened long-axis of the strain ellipse.

- C-planes or *cisaillement* planes form parallel to the shear zone boundary. The angle between the C and S planes is always acute, and defines the shear sense. Generally, the lower the C-S angle the greater the strain.

- The C' planes, also known as shear bands and secondary shear fabrics, are commonly observed in strongly foliated mylonites especially phyllonites, and form at an angle of about 20 degrees to the S-plane.

The sense of shear shown by both S-C and S-C' structures matches that of the shear zone in which they are found.

Other microstructures which can give sense of shear include:

- sigmoidal veins

- mica fish

- rotated porphyroclasts

- asymmetric boudins (Figure 1)

- asymmetric folds

Transpression

Transpression regimes are formed during oblique collision of tectonic plates and during non-orthogonal subduction. Typically a mixture of oblique-slip thrust faults and strike-slip or transform faults are formed. Microstructural evidence of transpressional regimes can be rodding lineations, mylonites, augen-structured gneisses, mica fish and so on.

A typical example of a transpression regime is the Alpine Fault zone of New Zealand, where the oblique subduction of the Pacific Plate under the Indo-Australian Plate is converted to oblique strike-slip movement. Here, the orogenic belt attains a trapezoidal shape dominated by oblique splay faults, steeply-dipping recumbent nappes and fault-bend folds.

The Alpine Schist of New Zealand is characterised by heavily crenulated and sheared phyllite. It is being pushed up at the rate of 8 to 10 mm per year, and the area is prone to large earthquakes with a south block up and west oblique sense of movement.

Transtension

Transtension regimes are oblique tensional environments. Oblique, normal geologic fault and detachment faults in rift zones are the typical structural manifestations of transtension conditions. Microstructural evidence of transtension includes rodding or stretching lineations, stretched porphyroblasts, mylonites, etc.

Saltation (Geology)

In geology, saltation (from Latin *saltus*, "leap") is a specific type of particle transport by fluids such as wind or water. It occurs when loose material is removed from a bed and carried by the fluid, before being transported back to the surface. Examples include pebble transport by rivers, sand drift over desert surfaces, soil blowing over fields, and snow drift over smooth surfaces such as those in the Arctic or Canadian Prairies.

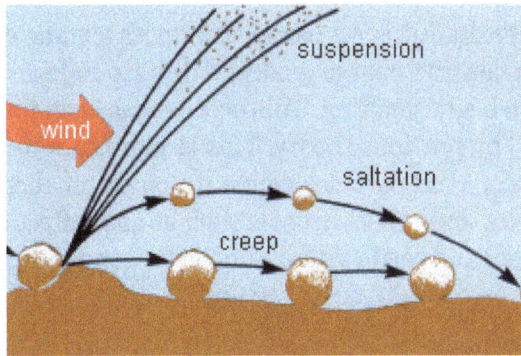

Saltation of sand

Saltation Process

At low fluid velocities, loose material rolls downstream, staying in contact with the surface. This is called *creep* or *reptation*. Here the forces exerted by the fluid on the particle are only enough to roll the particle around the point of contact with the surface.

Once the wind speed reaches a certain critical value, termed the *impact* or *fluid threshold*, the drag and lift forces exerted by the fluid are sufficient to lift some particles from the surface. These particles are accelerated by the fluid, and pulled downward by gravity, causing them to travel in roughly ballistic trajectories. If a particle has obtained sufficient speed from the acceleration by the fluid, it can eject, or *splash*, other particles in saltation, which propagates the process. Depending on the surface, the particle could also disintegrate on impact, or eject much finer sediment from the surface. In air, this process of *saltation bombardment* creates most of the dust in dust storms. In rivers, this process repeats continually, gradually eroding away the river bed, but also transporting-in fresh material from upstream.

Suspension generally affects small particles ('small' means ~70 micrometres or less for particles in air). For these particles, vertical drag forces due to turbulent fluctuations in the fluid are similar in magnitude to the weight of the particle. These smaller particles are carried by the fluid in suspension, and advected downstream. The smaller the particle, the less important the downward pull of gravity, and the longer the particle is likely to stay in suspension.

Saltating dune sand in a wind tunnel. (Photo credit: Wind Erosion Research Unit, USDA-ARS, Manhattan, Kansas)

A recent study finds that saltating sand particles induces a static electric field by friction. Saltating sand acquires a negative charge relative to the ground which in turn loosens more sand particles which then begin saltating. This process has been found to double the number of particles predicted by previous theory. This is significant in meteorology because it is primarily the saltation of sand particles which dislodges smaller dust particles into the atmosphere. Dust particles and other aerosols such as soot affect the amount of sunlight received by the atmosphere and earth, and are nuclei for condensation of the water vapour.

Avalanches

Saltation layers can also form in avalanches.

Aeolian Processes

Wind erosion of soil at the foot of Chimborazo, Ecuador.

Aeolian processes, also spelled eolian or æolian, pertain to wind activity in the study of geology and weather and specifically to the wind's ability to shape the surface of the Earth (or other planets). Winds may erode, transport, and deposit materials and are effective agents in regions with sparse vegetation, a lack of soil moisture and a large supply of unconsolidated sediments. Although water is a much more powerful eroding force than wind, aeolian processes are important in arid environments such as deserts.

Rock carved by drifting sand below Fortification Rock in Arizona
(Photo by Timothy H. O'Sullivan, USGS, 1871)

The term is derived from the name of the Greek god Aeolus, the keeper of the winds.

Wind Erosion

Wind erodes the Earth's surface by deflation (the removal of loose, fine-grained particles by the turbulent action of the wind) and by abrasion (the wearing down of surfaces by the grinding action and sandblasting of windborne particles).

A rock sculpted by wind erosion in the Altiplano region of Bolivia

Regions which experience intense and sustained erosion are called deflation zones. Most aeolian deflation zones are composed of desert pavement, a sheet-like surface of rock fragments that remains after wind and water have removed the fine particles. Almost half of Earth's desert surfaces are stony deflation zones. The rock mantle in desert pavements protects the underlying material from deflation.

Sand blowing off a crest in the Kelso Dunes of the Mojave Desert, California.

Wind-carved alcove in the Navajo Sandstone near Moab, Utah

A dark, shiny stain, called desert varnish or rock varnish, is often found on the surfaces of some desert rocks that have been exposed at the surface for a long period of time. Manganese, iron oxides, hydroxides, and clay minerals form most varnishes and provide the shine.

Deflation basins, called blowouts, are hollows formed by the removal of particles by wind. Blowouts are generally small, but may be up to several kilometers in diameter.

Wind-driven grains abrade landforms. In parts of Antarctica wind-blown snowflakes that are technically sediments have also caused abrasion of exposed rocks. Grinding by particles carried in the wind creates grooves or small depressions. Ventifacts are rocks which have been cut, and sometimes polished, by the abrasive action of wind.

Sculpted landforms, called yardangs, are up to tens of meters high and kilometers long and are forms that have been streamlined by desert winds. The famous Great Sphinx of Giza in Egypt may be a modified yardang.

Transport

Particles are transported by winds through suspension, saltation (skipping or bouncing) and creeping (rolling or sliding) along the ground.

Dust storm approaching Spearman, Texas April 14, 1935.

Dust storm in Amarillo, Texas. FSA photo by Arthur Rothstein (1936)

A massive sand storm cloud is about to envelop a military camp as it rolls over Al Asad, Iraq, just before nightfall on April 27, 2005.

Small particles may be held in the atmosphere in suspension. Upward currents of air support the weight of suspended particles and hold them indefinitely in the surrounding air. Typical winds near Earth's surface suspend particles less than 0.2 millimeters in diameter and scatter them aloft as dust or haze.

Saltation is downwind movement of particles in a series of jumps or skips. Saltation normally lifts sand-size particles no more than one centimeter above the ground and proceeds at one-half to one-third the speed of the wind. A saltating grain may hit other grains that jump up to continue the saltation. The grain may also hit larger grains that are too heavy to hop, but that slowly creep forward as they are pushed by saltating grains. Surface creep accounts for as much as 25 percent of grain movement in a desert.

Aeolian turbidity currents are better known as dust storms. Air over deserts is cooled significantly when rain passes through it. This cooler and denser air sinks toward the desert surface. When it reaches the ground, the air is deflected forward and sweeps up surface debris in its turbulence as a dust storm.

Crops, people, villages, and possibly even climates are affected by dust storms. Some dust storms are intercontinental, a few may circle the globe, and occasionally they may engulf entire planets. When the Mariner 9 spacecraft entered its orbit around Mars in 1971, a dust storm lasting one month covered the entire planet, thus delaying the task of photo-mapping the planet's surface.

Most of the dust carried by dust storms is in the form of silt-size particles. Deposits of this windblown silt are known as loess. The thickest known deposit of loess, 335 meters, is on the Loess Plateau in China. This very same Asian dust is blown for thousands of miles, forming deep beds in places as far away as Hawaii. In Europe and in the Americas, accumulations of loess are generally from 20 to 30 meters thick.

Aeolian transport from deserts plays an important role in ecosystems globally, e.g. by transport of minerals from the Sahara to Amazonia. Saharan dust is also responsible for forming red clay soils in southern Europe. Aeolian processes are affected by human activity, such as the use of 4x4 vehicles.

Small whirlwinds, called dust devils, are common in arid lands and are thought to be related to very intense local heating of the air that results in instabilities of the air mass. Dust devils may be as much as one kilometer high.

Deposition

Wind-deposited materials hold clues to past as well as to present wind directions and intensities. These features help us understand the present climate and the forces that molded it. Wind-deposited sand bodies occur as sand sheets, ripples, and dunes.

Sand sheets are flat, gently undulating sandy plots of sand surfaced by grains that may be too large for saltation. They form approximately 40 percent of aeolian depositional surfaces. The Selima Sand Sheet in the eastern Sahara Desert, which occupies 60,000 square kilometers in southern Egypt and northern Sudan, is one of the Earth's largest sand sheets. The Selima is absolutely flat in a few places; in others, active dunes move over its surface.

Cross-bedding of sandstone near Mount Carmel road, Zion National Park, indicating wind action and sand dune formation prior to formation of rock (NPS photo by George A. Grant, 1929)

Mesquite Flat Dunes in Death Valley looking toward the Cottonwood Mountains from the north west arm of Star Dune (2003)

Holocene eolianite deposit on Long Island, The Bahamas. This unit is formed of wind-blown carbonate grains. (2007)

Wind blowing on a sand surface ripples the surface into crests and troughs whose long axes are perpendicular to the wind direction. The average length of jumps during saltation corresponds to the wavelength, or distance between adjacent crests, of the ripples. In ripples, the coarsest materials collect at the crests causing inverse grading. This distinguishes small ripples from dunes, where the coarsest materials are generally in the troughs. This is also a distinguishing feature between water laid ripples and aeolian ripples.

Wind-blown sand moves up the gentle upwind side of the dune by saltation or creep. Sand accumulates at the brink, the top of the slipface. When the buildup of sand at the brink exceeds the angle of repose, a small avalanche of grains slides down the slipface. Grain by grain, the dune moves downwind.

Accumulations of sediment blown by the wind into a mound or ridge, dunes have gentle upwind slopes on the windward side. The downwind portion of the dune, the lee slope, is commonly a steep avalanche slope referred to as a slipface. Dunes may have more than one slipface. The minimum height of a slipface is about 30 centimeters.

Some of the most significant experimental measurements on aeolian sand movement were performed by Ralph Alger Bagnold, a British engineer who worked in Egypt prior to World War II. Bagnold investigated the physics of particles moving through the atmosphere and deposited by wind. He recognized two basic dune types, the crescentic dune, which he called "barchan," and the linear dune, which he called longitudinal or "seif" (Arabic for "sword").

A 2011 study published in *Catena* examined the effect of vegetation on aeolian dust accumulation in the semiarid steppe of northern China. Using a series of trays with different vegetation coverage and a control model with none, the authors found that an increase in vegetation coverage improves the efficiency of dust accumulation and adds more nutrients to the environment, particularly organic carbon. Two critical point were revealed by their data: 1. the efficiency of trapping dust increases slowly above 15% coverage, and decreases rapidly below 15% coverage. 2. at around 55%-75% coverage, dust accumulation reaches a maximum capacity.

A three year quantitative study on the effects of vegetation removal on wind erosion found that the removal of grasses in an aeolian environment increased the rate of soil deposition. In the same study, a relationship was shown between decreasing plant density with decreasing soil nutrients. Similarly, horizontal soil flux across the test site was shown to increase with increasing vegetation removal.

A 1998 study published in Earth Surfaces Processes and Landforms investigated the relationship between vegetative cover on sand surfaces with the rate of sand transport. It was found that sand flux decreased exponentially with vegetation cover. This was done by measuring plots of land with varying degrees of vegetation against rates of sand transport. The authors contend that this relationship can be utilized to manipulate rates of sediment flux by introducing vegetation in an area or to quantify human impact by recognizing vegetation loss's affect on sandy landscapes.

Igneous Differentiation

In geology, igneous differentiation is an umbrella term for the various processes by which magmas undergo bulk chemical change during the partial melting process, cooling, emplacement, or eruption.

Definitions

Primary Melts

When a rock melts to form a liquid, the liquid is known as a *primary melt*. Primary melts have not undergone any differentiation and represent the starting composition of a magma. In nature, primary melts are rarely seen. Some leucosomes of migmatites are examples of primary melts. Primary melts derived from the mantle are especially important and are known as *primitive melts* or primitive magmas. By finding the primitive magma composition of a magma series, it is possible to model the composition of the rock from which a melt was formed, which is important because we have little direct evidence of the Earth's mantle.

Parental Melts

Where it is impossible to find the primitive or primary magma composition, it is often useful to attempt to identify a parental melt. A parental melt is a magma composition from which the observed range of magma chemistries has been derived by the processes of igneous differentiation. It need not be a primitive melt.

For instance, a series of basalt lava flows is assumed to be related to one another. A composition from which they could reasonably be produced by fractional crystallization is termed a *parental melt*. To prove this, fractional crystallization models would be produced to test the hypothesis that they share a common parental melt.

Cumulate Rocks

Fractional crystallization and accumulation of crystals formed during the differentiation process of a magmatic event are known as *cumulate rocks*, and those parts are the first which crystallize out of the magma. Identifying whether a rock is a cumulate or not is crucial for understanding if it can be modelled back to a primary melt or a primitive melt, and identifying whether the magma has dropped out cumulate minerals is equally important even for rocks which carry no phenocrysts.

Underlying Causes of Differentiation

The primary cause of change in the composition of a magma is *cooling*, which is an inevitable consequence of the magma being created and migrating from the site of partial melting into an area of lower stress - generally a cooler volume of the crust.

Cooling causes the magma to begin to crystallize minerals from the melt or liquid portion of the magma. Most magmas are a mixture of liquid rock (melt) and minerals (phenocrysts).

Contamination is another cause of magma differentiation. Contamination can be caused by *assimilation* of wall rocks, mixing of two or more magmas or even by replenishment of the magma chamber with fresh, hot magma.

The whole gamut of mechanisms for differentiation has been referred to as the FARM process, which stands for Fractional crystallization, Assimilation, Replenishment and Magma mixing.

Fractional Crystallization of Igneous Rocks

Fractional crystallization is the removal and segregation from a melt of mineral precipitates, which changes the composition of the melt. This is one of the most important geochemical and physical processes operating within the Earth's crust and mantle.

Fractional crystallization in silicate melts (magmas) is a very complex process compared to chemical systems in the laboratory because it is affected by a wide variety of phenomena. Prime amongst these are the composition, temperature, and pressure of a magma during its cooling.

The composition of a magma is the primary control on which mineral is crystallized as the melt cools down past the liquidus. For instance in mafic and ultramafic melts, the MgO and SiO_2 contents determine whether forsterite olivine is precipitated or whether enstatite pyroxene is precipitated.

Two magmas of similar composition and temperature at different pressure may crystallize different minerals. An example is high-pressure and high-temperature fractional crystallization of granites to produce single-feldspar granite, and low-pressure low-temperature conditions which produce two-feldspar granites.

The partial pressure of volatile phases in silicate melts is also of prime importance, especially in near-solidus crystallization of granites.

Assimilation

Assimilation is a popular mechanism for explaining the felsification of ultramafic and mafic magmas as they rise through the crust. Assimilation assumes that a hot primitive melt intruding into a cooler, felsic crust will melt the crust and mix with the resulting melt. This then alters the composition of the primitive magma.

Effects of this kind are to be expected, and have been clearly proven in many places. There is, however, a general reluctance to admit that they are of great importance. The nature and succession of the rock types do not as a rule show any relation to the sedimentary or other materials which may be supposed to have been dissolved; and where solution is known to have gone on the products are usually of abnormal character and easily distinguishable from the common rock types.

Replenishment

When a melt undergoes cooling along the liquid line of descent, the results are limited to the production of a homogeneous solid body of intrusive rock, with uniform mineralogy and composition, or a partially differentiated cumulate mass with layers, compositional zones and so on. This behaviour is fairly predictable and easy enough to prove with geochemical investigations. In such cases, a magma chamber will form a close approximation of the ideal Bowen's reaction series. However, most magmatic systems are polyphase events, with several pulses of magmatism. In such a case, the liquid line of descent is interrupted by the injection of a fresh batch of hot, undifferentiated magma. This can cause extreme fractional crystallisation because of three main effects:

- Additional heat provides additional energy to allow more vigorous convection, allows resorption of existing mineral phases back into the melt, and can cause a higher-temperature form of a mineral or other higher-temperature minerals to begin precipitating

- Fresh magma changes the composition of the melt, changing the chemistry of the phases which are being precipitated. For instance, plagioclase conforms to the liquid line of descent by forming initial anorthite which, if removed, changes the equilibrium mineral composition to oligoclase or albite. Replenishment of the magma can see this trend reversed, so that more anorthite is precipitated atop cumulate layers of albite.

- Fresh magma destabilises minerals which are precipitating as solid solution series or on a eutectic; a change in composition and temperature can cause extremely rapid crystallisation of certain mineral phases which are undergoing a eutectic crystallisation phase.

Magma Mixing

Magma mixing is the process by which two magmas meet, comingle, and form a magma of a composition somewhere between the two end-member magmas.

Magma mixing is a common process in volcanic magma chambers, which are open-system chambers where magmas enter the chamber, undergo some form of assimilation, fractional crystallisation and partial melt extraction (via eruption of lava), and are replenished.

Magma mixing also tends to occur at deeper levels in the crust and is considered one of the primary mechanisms for forming intermediate rocks such as monzonite and andesite. Here, due to heat transfer and increased volatile flux from subduction, the silicic crust melts to form a felsic magma (essentially granitic in composition). These granitic melts are known as an *underplate*. Basaltic primary melts formed in the mantle beneath the crust rise and mingle with the underplate magmas, the result being part-way between basalt and rhyolite; literally an 'intermediate' composition.

Other Mechanisms of Differentiation

Interface entrapment Convection in a large magma chamber is subject to the interplay of forces generated by thermal convection and the resistance offered by friction, viscosity and drag on the magma offered by the walls of the magma chamber. Often near the margins of a magma chamber which is convecting, cooler and more viscous layers form concentrically from the outside in, defined by breaks in viscosity and temperature. This forms laminar flow, which separates several domains of the magma chamber which can begin to differentiate separately.

Flow banding is the result of a process of fractional crystallization which occurs by convection, if the crystals which are caught in the flow-banded margins are removed from the melt. The friction and viscosity of the magma causes phenocrysts and xenoliths within the magma or lava to slow down near the interface and become trapped in a viscous layer. This can change the composition of the melt in large intrusions, leading to differentiation.

Partial Melt Extraction

With reference to the definitions, above, a magma chamber will tend to cool down and crystallize minerals according to the liquid line of descent. When this occurs, especially in conjunction with zonation and crystal accumulation, and the melt portion is removed, this can change the composition of a magma chamber. In fact, this is basically fractional crystallization, except in this case we are observing a magma chamber which is the remnant left behind from which a daughter melt has been extracted.

If such a magma chamber continues to cool, the minerals it forms and its overall composition will not match a sample liquid line of descent or a parental magma composition.

Typical Behaviours of Magma Chambers

It is worth reiterating that magma chambers are not usually static single entities. The typical magma chamber is formed from a series of injections of melt and magma, and most are also subject to some form of partial melt extraction.

Granite magmas are generally much more viscous than mafic magmas and are usually more homogeneous in composition. This is generally considered to be caused by the viscosity of the magma, which is orders of magnitude higher than mafic magmas. The higher viscosity means that, when melted, a granitic magma will tend to move in a larger concerted mass and be emplaced as a larger mass because it is less fluid and able to move. This is why granites tend to occur as large plutons, and mafic rocks as dikes and sills.

Granites are cooler and are therefore less able to melt and assimilate country rocks. Wholesale contamination is therefore minor and unusual, although mixing of granitic and basaltic melts is not unknown where basalt is injected into granitic magma chambers.

Mafic magmas are more liable to flow, and are therefore more likely to undergo periodic replenishment of a magma chamber. Because they are more fluid, crystal precipitation occurs much more rapidly, resulting in greater changes by fractional crystallisation. Higher temperatures also allow mafic magmas to assimilate wall rocks more readily and therefore contamination is more common and better developed.

Dissolved Gases

All igneous magmas contain dissolved gases (water, carbonic acid, hydrogen sulfide, chlorine, fluorine, boric acid, etc.). Of these water is the principal, and was formerly believed to have percolated downwards from the Earth's surface to the heated rocks below, but is now generally admitted to be an integral part of the magma. Many peculiarities of the structure of the plutonic rocks as contrasted with the lavas may reasonably be accounted for by the operation of these gases, which were unable to escape as the deep-seated masses slowly cooled, while they were promptly given up by the superficial effusions. The acid plutonic or intrusive rocks have never been reproduced by laboratory experiments, and the only successful attempts to obtain their minerals artificially have been those in which special provision was made for the retention of the "mineralizing" gases in the crucibles or sealed tubes employed. These gases often do not enter into the composition of the rock-forming minerals, for most of these are free from water, carbonic acid, etc. Hence as crystallization goes on the residual melt must contain an ever-increasing proportion of volatile constituents. It is conceivable that in the final stages the still uncrystallized part of the magma has more resemblance to a solution of mineral matter in superheated steam than to a dry igneous fusion. Quartz, for example, is the last mineral to form in a granite. It bears much of the stamp of the quartz which we know has been deposited from aqueous solution in veins, etc. It is at the same time the most infusible of all the common minerals of rocks. Its late formation shows that in this case it arose at comparatively low temperatures and points clearly to the special importance of the gases of the magma as determining the sequence of crystallization.

When solidification is nearly complete the gases can no longer be retained in the rock and make their escape through fissures towards the surface. They are powerful agents in attacking the minerals of the rocks which they traverse, and instances of their operation are found in the kaolinization of granites, tourmalinization and formation of greisen, deposition of quartz veins, and the group of changes known as propylitization. These "pneumatolytic" processes are of the first importance in the genesis of many ore deposits. They are a real part of the history of the magma itself and constitute the terminal phases of the volcanic sequence.

Quantifying Igneous Differentiation

There are several methods of directly measuring and quantifying igneous differentiation processes;

- Whole rock geochemistry of representative samples, to track changes and evolution of the magma systems

    o Using the above, calculating normative mineralogy and investigating trends

- Trace element geochemistry

- Isotope geochemistry

    o Investigating the contamination of magma systems by wall rock assimilation using radiogenic isotopes

In all cases, the primary and most valuable method for identifying magma differentiation processes is mapping the exposed rocks, tracking mineralogical changes within the igneous rocks and describing field relationships and textural evidence for magma differentiation.

Metasomatism

Metasomatism is the chemical alteration of a rock by hydrothermal and other fluids. It is the replacement of one rock by another of different minerological and chemical composition. The minerals which compose the rocks are dissolved and new mineral formations are deposited in their place. Dissolution and deposition occur simultaneously and the rock remains solid.

Synonyms to the word metasomatism are metasomatose and metasomatic process. The word metasomatose can also be used as a name for specific varieties of metasomatism (for example *Mg-metasomatose* and *Na-metasomatose*).

Metasomatism can occur via the action of hydrothermal fluids from an igneous or metamorphic source.

In the igneous environment, metasomatism creates skarns, greisen, and may affect hornfels in the contact metamorphic aureole adjacent to an intrusive rock mass. In the metamorphic environment, metasomatism is created by mass transfer from a volume of metamorphic rock at higher stress and temperature into a zone with lower stress and temperature, with metamorphic hydrothermal solutions acting as a solvent. This can be envisaged as the metamorphic rocks within the deep crust losing fluids and dissolved mineral components as hydrous minerals break down, with this fluid percolating up into the shallow levels of the crust to chemically change and alter these rocks.

Metasomatic albite + hornblende + tourmaline alteration of metamorphosed granite, Stone Mountain, Atlanta

This mechanism implies that metasomatism is open system behaviour, which is different from classical metamorphism which is the in-situ mineralogical change of a rock without appreciable change in the chemistry of the rock. Because metamorphism usually requires water in order to facilitate metamorphic reactions, metasomatism and metamorphism nearly always occur together.

Further, because metasomatism is a mass transfer process, it is not restricted to the rocks which are changed by addition of chemical elements and minerals or hydrous compounds. In all cases, to produce a metasomatic rock some other rock is also metasomatised, if only by *dehydration* reactions with minimal chemical change. This is best illustrated by gold ore deposits which are the product of focused concentration of fluids derived from many cubic kilometres of dehydrated crust into thin, often highly metasomatised and altered shear zones and lodes. The source region is often largely chemically unaffected compared to the highly hydrated, altered shear zones, but both must have undergone complementary metasomatism.

Metasomatism is more complicated in the Earth's mantle, because the composition of peridotite at high temperatures can be changed by infiltration of carbonate and silicate melts and by carbon dioxide-rich and water-rich fluids, as discussed by Luth (2003). Metasomatism is thought to be particularly important in changing the composition of mantle peridotite below island arcs as water is driven out of ocean lithosphere during subduction. Metasomatism has also been considered critical for enriching source regions of some silica-undersaturated magmas. Carbonatite melts are often considered to have been responsible for enrichment of mantle peridotite in incompatible elements.

Types of Metasomatites

Metasomatic rocks can be extremely varied. Often, metasomatised rocks are pervasively but weakly *altered*, such that the only evidence of alteration is bleaching, change in colour or change in the crystallinity of micaceous minerals.

In such cases, characterising alteration often requires microscope investigation of the mineral assemblage of the rocks to characterise the minerals, any additional mineral growth, changes in protolith minerals, and so on.

In some cases, geochemical evidence can be found of metasomatic alteration processes. This is usually in the form of mobile, soluble elements such as barium, strontium, rubidium, calcium and some rare earth elements. However, to characterise the alteration properly, it is necessary to compare altered with unaltered samples.

When the process becomes extremely advanced, typical metasomatites can include:

- Chlorite or mica whole-rock replacement in shear zones, resulting in rocks in which the existing mineralogy has been completely recrystallised and replaced by hydrated minerals such as chlorite, muscovite, and serpentine.

- Skarn and skarnoid rock types, typically adjacent to granite intrusions and adjacent to reactive lithologies such as limestone, marl and banded iron formation.

- Greisen deposits within granite margins and cupolas.

Effects of metasomatism in mantle peridotite can be either modal or cryptic. In cryptic metasomatism, mineral compositions are changed, or introduced elements are concentrated on grain boundaries and the peridotite mineralogy appears unchanged. In modal metasomatism, new minerals are formed.

Cryptic metasomatism may be caused as rising or percolating melts interact with surrounding peridotite, and compositions of both melts and peridotite are changed. At high mantle temperatures, solid-state diffusion can also be effective in changing rock compositions over tens of centimeters adjacent to melt conduits: gradients in mineral composition adjacent to pyroxenite dikes may preserve evidence of the process.

Modal metasomatism may result in formation of amphibole and phlogopite, and the presence of these minerals in peridotite xenoliths has been considered strong evidence of metasomatic processes in the mantle. Formation of minerals less common in peridotite, such as dolomite, calcite, ilmenite, rutile, and armalcolite, is also attributed to melt or fluid metasomatism.

Alteration Assemblages

Investigation of altered rocks in hydrothermal ore deposits has highlghted several ubiquitous types of *alteration assemblages* which create distinct groups of metasomatic alteration effects, textures and mineral assemblages.

- Propylitic alteration is caused by iron and sulfur-bearing hydrothermal fluids, and typically results in epidote-chlorite-pyrite alteration, often with hematite and magnetite facies

- Albite-epidote alteration is caused by silica-bearing fluids rich in sodium and calcium, and typically results in weak albite-silica-epidote

- Potassic alteration, typical of porphyry copper and lode gold deposits, results in production of micaceous, potassic minerals such as biotite in iron-rich rocks, muscovite mica or sericite in felsic rocks, and orthoclase (adularia) alteration, often quite pervasive and producing distinct salmon-pink alteration vein selvages.

Rarer types of hydrothermal fluids may include highly carbonic fluids, resulting in advanced carbonation reactions of the host rock typical of calc-silicates, and silica-hematite fluids resulting in production of jasperoids, manto ore deposits and pervasive zones of silicification, typically in dolomite formations. Stressed minerals and country rocks of granitic plutons are replaced by porphyroblasts of orthoclase and quartz, in Papoose Flat quartz monzonites Dickson, (1996, 2000, 2005).

Exfoliation Joint

Exfoliation joints or sheet joints are surface-parallel fracture systems in rock often leading to erosion of concentric slabs.

Exfoliation joints wrapping around Half Dome in Yosemite National Park, California.

Exfoliation joints in granite at Enchanted Rock State Natural Area, Texas, USA. Detached blocks have slid along the steeply-dipping joint plane.

General Characteristics of Exfoliation Joints

Small scale example of exfoliation jointing on a dolerite dyke, Pilbara, Western Australia

- Commonly follow topography.

- Divide the rock into sub-planar slabs.

- Joint spacing increases with depth from a few centimeters near the surface to a few meters

- Maximum depth of observed occurrence is around 100 meters.

- Deeper joints have a larger radius of curvature, which tends to round the corners of the landscape as material is eroded

- Fracture mode is tensile

- Occur in many different lithologies and climate zones, not unique to glaciated landscapes.

- Host rock is generally sparsely jointed, fairly isotropic, and has high compressive strength.

- Can have concave and convex upwards curvatures.

- Often associated with secondary compressive forms such as arching, buckling, and A-tents (buckled slabs)

Formation of Exfoliation Joints

Despite their common occurrence in many different landscapes, geologists have yet to reach an agreement on a general theory of exfoliation joint formation. Many different theories have been suggested, below is a short overview of the most common.

Removal of Overburden and Rebound

This theory was originally proposed by the pioneering geomorphologist Grove Karl Gilbert in 1904 and is widely found in introductory geology texts. The basis of this theory is that erosion of overburden and exhumation of deeply buried rock to the ground surface allows previously compressed rock to expand radially, creating tensile stress and fracturing the rock in layers parallel to the ground surface. The description of this mechanism has led to alternate terms for exfoliation joints, including pressure release or offloading joints. Though the logic of this theory is appealing, there are many inconsistencies with field and laboratory observations suggesting that it may be incomplete, such as:

- Exfoliation joints can be found in rocks that have never been deeply buried.

- Laboratory studies show that simple compression and relaxation of rock samples under realistic conditions does not cause fracturing.

- Exfoliation joints are most commonly found in regions of surface-parallel compressive stress, whereas this theory calls for them to occur in zones of extension.

Exfoliation joints exposed in a road cut in Yosemite National Park, California.

One possible extension of this theory to match with the *compressive stress* theory (outlined below) is as follows (Goodman, 1989): The exhumation of deeply buried rocks relieves vertical stress, but horizontal stresses can remain in a competent rock mass since the medium is laterally confined. Horizontal stresses become aligned with the current ground surface as the vertical stress drops to zero at this boundary. Thus large surface-parallel compressive stresses can be generated through exhumation that may lead to tensile rock fracture as described below.

Thermoelastic Strain

Rock expands upon heating and contracts upon cooling and different rock-forming minerals have variable rates of thermal expansion / contraction. Daily rock surface temperature variations can be quite large, and many have suggested that stresses created during heating cause the near-surface zone of rock to expand and detach in thin slabs (e.g. Wolters, 1969). Large diurnal or fire-induced temperature fluctuations have been observed to create thin lamination and flaking at the surface of rocks, sometimes labeled exfoliation. However, since diurnal temperature fluctuations only reach a few centimeters depth in rock (due to rock's low thermal conductivity), this theory cannot account for the observed depth of exfoliation jointing that may reach 100 meters.

Chemical Weathering

Mineral weathering by penetrating water can cause flaking of thin shells of rock since the volume of some minerals increases upon hydration. However, not all mineral hydration results in increased volume, while field observations of exfoliation joints show that the joint surfaces have not experienced significant chemical alteration, so this theory can be rejected as an explanation for the origin of large-scale, deeper exfoliation joints.

Compressive Stress and Extensional Fracture

Exfoliation joints have modified the near-surface portions of massive granitic rocks in Yosemite National Park, helping create the many spectacular domes, including Half Dome shown here.

Large compressive stresses parallel to the land (or a free) surface can create tensile mode fractures in rock, where the direction of fracture propagation is parallel to the greatest principle compressive stress and the direction of fracture opening is perpendicular to the free surface. This type of fracturing has been observed in the laboratory since at least 1900 (in both uniaxial and biaxial unconfined compressive loading. Tensile cracks can form in a compressive stress field due to the influence of pervasive microcracks in the rock lattice and extension of so-called *wing cracks* from near the tips of preferentially oriented microcracks, which then curve and align with the direction of the principle compressive stress. Fractures formed in this way are sometimes called axial cleavage, longitudinal splitting, or extensional fractures, and are commonly observed in the laboratory during uniaxial compression tests. High horizontal or surface-parallel compressive stress can result from regional tectonic or topographic stresses, or by erosion or excavation of overburden.

With consideration of the field evidence and observations of occurrence, fracture mode, and secondary forms, high surface-parallel compressive stresses and extensional fracturing (axial cleavage) seems to be the most plausible theory explaining the formation of exfoliation joints.

Engineering Geology Significance

Recognizing the presence of exfoliation joints can have important implications in geological engineering. Most notable may be their influence on slope stability. Exfoliation joints following the topography of inclined valley walls, bedrock hill slopes, and cliffs can create rock blocks that are particularly prone to sliding. Especially when the toe of the slope is undercut (naturally or by human activity), sliding along exfoliation joint planes is likely if the joint dip exceeds the joint's frictional angle. Foundation work may also be affected by the presence of exfoliation joints, for example in the case of dams. Exfoliation joints underlying a dam foundation can create a significant leakage hazard, while increased water pressure in joints may result in lifting or sliding of the dam. Finally, exfoliation joints can exert strong directional control on groundwater flow and contaminant transport.

Denudation

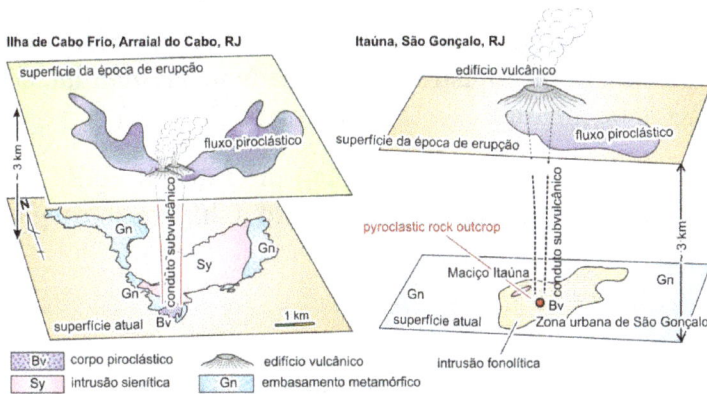

Schematic illustration of regional denudation for felsic alkaline intrusive rock bodies of the State of Rio de Janeiro, Brazil: Cabo Frio Island and Itaúna Body.

In geology, denudation is the long-term sum of processes that cause the wearing away of the Earth's surface by moving water, ice, wind and waves, leading to a reduction in elevation and relief of landforms and landscapes. Endogenous processes such as volcanoes, earthquakes, and plate tectonics uplift and expose continental crust to the exogenous processes of weathering, erosion, and mass wasting.

Processes

Denudation incorporates the mechanical, biological and chemical processes of erosion, weathering and mass wasting. Denudation can involve the removal of both solid particles and dissolved material. These include sub-processes of cryofracture, insolation weathering, slaking, salt weathering, bioturbation and anthropogenic impacts.

Factors affecting denudation include:´

- Anthropogenic activity

- Biosphere

- Climate (most directly in chemical weathering)

- Geology

- Surface topography

- Tectonic activity

Rates

Modern denudation estimates are usually based on stream load measurements taken at gauging stations. Suspended load, bed load, and dissolved load are included in measurements. The weight of the load is converted to volumetric units and the load volume is divided by the area of the watershed above the gaging station. The result is an estimate of the wearing down of the Earth's surface in inches or centimeters per 1000 years. In most cases no adjustments are made for human impact, which causes the measurements to be inflated.

Denudation rates are usually much lower than the rates of uplift. The only areas at which there could be equal rates of denudation and uplift are active plate margins with an extended period of continuous deformation.

Proposed Cycles

Early studies prompted the formation of denudation cycle hypotheses to describe land formations. Although at present they are mostly discounted, many of these models are enduring due to their simplicity and seemingly obvious assumptions.

In the 1890s W. M. Davis proposed a cycle of 'wearing down' in which so called 'young' landscapes had high gradients and elevations, and waning, low elevation topography through middle age to old age. Landscapes of Britain and Wales were thought to reflect these multiple peneplanation and rejuvenation cycles, such as the 3,000-foot remnant summit plateau in North Wales. A number of assumptions of fluvial and glacial dynamics in temperate areas were made in the formation of this model.

Such theories were proposed before tectonic theory was largely understood, and therefore are now largely discredited.

Volcanic Landforms

Denudation exposes deep sub-volcanic structures on the present surface of the area where volcanic activity once occurred. Sub-volcanic structures such as neck and dyke (volcanic vent) are exposed by denudation.

A) Villarica Volcano, Chile, a volcano without effects of erosion and denudation
B) Chachahén Volcano, Mendoza, Argentina, a volcano with strong effect of erosion but no denudation
C) Cardiel Lake, Santa Cruz, Argentina, a volcanic area under strong effect of denudation, exposing subvolcanic rock body.

Spheroidal Weathering

Spheroidal or woolsack weathering in granite on Haytor, Dartmoor, England

Spheroidal weathering in granite, Estaca de Bares, A Coruña, Galicia, Spain).

Spheroidal weathering is a form of chemical weathering that affects jointed bedrock and results in the formation of concentric or spherical layers of highly decayed rock within weathered bedrock that is known as *saprolite*. When saprolite is exposed by physical erosion, these concentric layers peel (spall) off as concentric shells much like the layers of a peeled onion. Within saprolite, spheroidal weathering often creates rounded boulders, known as *corestones* or *woolsack,* of relatively unweathered rock. Spheroidal weathering is also called onion skin weathering, concentric weathering, spherical weathering, or woolsack weathering.

Weathering Process

Spheroidal weathering is the result of chemical weathering of systematically jointed, massive rocks, including granite, dolerite, basalt and sedimentary rocks such as silicified sandstone. It occurs as the result of the chemical alteration of such rocks along intersecting joints. The chemical alteration of the rock results in the formation of abundant secondary minerals such as kaolinite, sericite, serpentine, montmorillonite, and chlorite and a corresponding increase in the volume of the altered rock. When the joints within bedrock form a 3-dimensional network, they subdivided it into separated blocks, often in the form of cubes or rectangles, that are bounded by these joints. Because water can penetrate the bedrock along these joints, the near-surface bedrock will be altered by weathering progressively inward along the faces of these blocks. The alteration by weathering of the bedrock will be greatest along the corners of each block, followed by the edges, and finally the faces of the cube. The differences in weathering rates between the corners, edges, and faces of a bedrock block will result in the formation of spheroidal layers of altered rock that surround an unaltered rounded boulder-size core of relatively unaltered rock known as a *corestone* or *woolsack*. Spheroidal weathering has often been incorrectly attributed solely to various types of physical weathering.

Frequently, erosion has removed the layers of altered rock and other saprolite surrounding corestones that were produced by spheroidal weathering. This leaves many corestones as freestanding boulders on the ground's surface. Often the spheroidal weathering, which created these corestones and the enclosing saprolite occurred in the prehistoric past during periods of humid, even tropical climates. Frequently, the removal of the saprolite by erosion and exposure of corestones as freestanding residual boulders, tors, or other landforms occurs many thousands of years later and during vastly different climatic conditions.

Depending on local environmental conditions, spheroidal weathering of bedrock blocks defined by tectonically induced joints and fractures may result in the formation of prominent and well-defined Liesegang rings within these blocks. These blocks typically consist of bedrock blocks (*Liesegang blocks*), which are bounded on their periphery by joints and fractures, and, in sedimentary rocks, bedding planes above and below. Each Liesegang block consists of a relatively unaltered core surrounded by concentric, alternating shells of iron-poor (intermediate shells) and iron-rich ('iron' shells) composition which make up the Liesegang rings. These iron-poor and iron-rich shells follow the configuration of the outer shape of the block and are sub-parallel to its sides. The iron-rich and iron-poor shells vary in degree of cementation and, as a result, can produce *box work* weathering structures during subsequent erosion. The degree of development of Liesegang rings as the result of weathering depends upon the spacing of the joint systems, groundwater flow, local topography, bedrock composition, and bed thickness.

References

- Sudipta Sengupta; Subir Kumar Ghosh; Kshitindramohan Naha (1997). Evolution of geological structures in micro- to macro-scales. Springer. p. 222. ISBN 0-412-75030-9.

- Neville J. Price; John W. Cosgrove (1990). "Figure 10.14: Classification of fold profiles using dip isogon patterns". Analysis of geological structures. Cambridge University Press. p. 246. ISBN 0-521-31958-7.

- Foundations of structural geology (3rd ed.). Routledge. p. 31 ff. ISBN 0-7487-5802-X.

- Ramsay, J.G.; Huber M.I. (1987). The techniques of modern structural geology. 2 (3 ed.). Academic Press. p. 392. ISBN 978-0-12-576922-8. Retrieved 2009-11-01.

- Nichols, G. (1999). "17. Sediments into rocks: post-depositional processes". Sedimentology and stratigraphy. Wiley-Blackwell. p. 355. ISBN 978-0-632-03578-6. Retrieved 2009-10-31.

- Hyne, N.J. (2001). Nontechnical guide to petroleum geology, exploration, drilling, and production. PennWell Books. p. 598. ISBN 978-0-87814-823-3. Retrieved 2009-11-01.

- Arvid M. Johnson; Raymond C. Fletcher (1994). "Figure 2.6". Folding of viscous layers: mechanical analysis and interpretation of structures in deformed rock. Columbia University Press. p. 87. ISBN 0-231-08484-6.

- Park, R.G. (1997). Foundations of structural geology (3rd ed.). Routledge. p. 109. ISBN 0-7487-5802-X.; RJ Twiss; EM Moores (1992). "Figure 12.8: Passive shear folding". Structural geology (2nd ed.). Macmillan. pp. 241–242. ISBN 0-7167-2252-6.

- Twidale, CR, and JRV Romani (2005) Landforms and Geology of Granite Terrains. A.A. Balkema Publishers Leiden, The Netherlands. 330 pp. ISBN 0-415-36435-3

- Migon, P (2006) Granite Landscapes of the World. (Geomorphological Landscapes of the World) Oxford University Press Inc., New York. 384 pp. ISBN 0-19-927368-5

- DD Pollard; RC Fletcher (2005). "Figure 3.14: Geometric attributes of folded geological surfaces". Fundamentals of Structural Geology. Cambridge University Press. p. 92. ISBN 0-521-83927-0.

Geology: Mapping and Modelling

Geological mapping is a vital part of geology; it is a special-purpose map that is created to illustrate geological features. The rocks or geological strata demonstrated in the map is shown by different colors and symbols. The major components of geology are discussed in the following text.

Geologic Map

A geologic map or geological map is a special-purpose map made to show geological features. Rock units or geologic strata are shown by color or symbols to indicate where they are exposed at the surface. Bedding planes and structural features such as faults, folds, foliations, and lineations are shown with strike and dip or trend and plunge symbols which give these features' three-dimensional orientations.

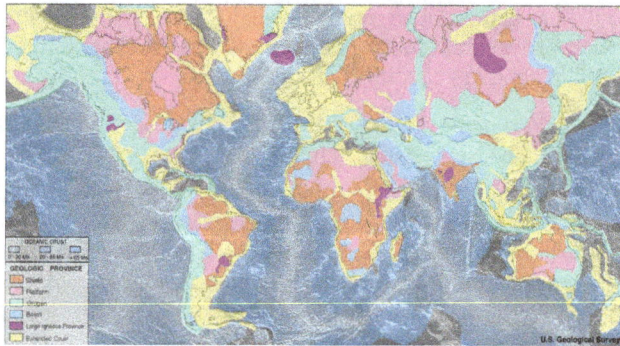

Mapped global geologic provinces

Stratigraphic contour lines may be used to illustrate the surface of a selected stratum illustrating the subsurface topographic trends of the strata. Isopach maps detail the variations in thickness of stratigraphic units. It is not always possible to properly show this when the strata are extremely fractured, mixed, in some discontinuities, or where they are otherwise disturbed.

Symbols

Lithologies

Rock units are typically represented by colors. Instead of (or in addition to) colors, certain symbols can be used. Different geologic mapping agencies and authorities have different standards for the colors and symbols to be used for rocks of differing types and ages.

Orientations

Geologists take two major types of orientation measurements (using a hand compass like a Brunton compass): orientations of planes and orientations of lines. Orientations of planes are often measured as a "strike" and "dip", while orientations of lines are often measured as a "trend" and "plunge".

A standard Brunton Geological compass, used commonly by geologists

Strike and dip symbols consist of a long "strike" line, which is perpendicular to the direction of greatest slope along the surface of the bed, and a shorter "dip" line on side of the strike line where the bed is going downwards. The angle that the bed makes with the horizontal, along the dip direction, is written next to the dip line. In the azimuthal system, strike and dip are often given as "strike/dip" (for example: 270/15, for a strike of west and a dip of 15 degrees below the horizontal).

Trend and plunge are used for linear features, and their symbol is a single arrow on the map. The arrow is oriented in the downgoing direction of the linear feature (the "trend") and at the end of the arrow, the number of degrees that the feature lies below the horizontal (the "plunge") is noted. Trend and plunge are often notated as PLUNGE → TREND (for example: 34 → 86 indicates a feature that is angled at 34 degrees below the horizontal at an angle that is just East of true South).

History

The oldest preserved geologic map is the Turin papyrus (1150 BCE), which shows the location of building stone and gold deposits in Egypt.

The earliest geologic map of the modern era is the 1771 "Map of Part of Auvergne, or figures of, The Current of Lava in which Prisms, Balls, Etc. are Made from Basalt. To be used with Mr. Demarest's theories of this hard basalt. Engraved by Messr. Pasumot and Daily, Geological Engineers of the King." This map is based on Nicolas Desmarest's

1768 detailed study of the geology and eruptive history of the Auvergne volcanoes and a comparison with the columns of the Giant's Causeway of Ireland. He identified both landmarks as features of extinct volcanoes. The 1798 report was incorporated in the 1771 (French) Royal Academy of Science compendium.

The first geological map of the U.S. was produced in 1809 by William Maclure. In 1807, Maclure undertook the self-imposed task of making a geological survey of the United States. He traversed and mapped nearly every state in the Union. During the rigorous two-year period of his survey, he crossed and recrossed the Allegheny Mountains some 50 times. Maclure's map shows the distribution of five classes of rock in what are now only the eastern states of the present-day US.

The first geologic map of Great Britain was created by William Smith in 1815.

Maps and Mapping Around the Globe

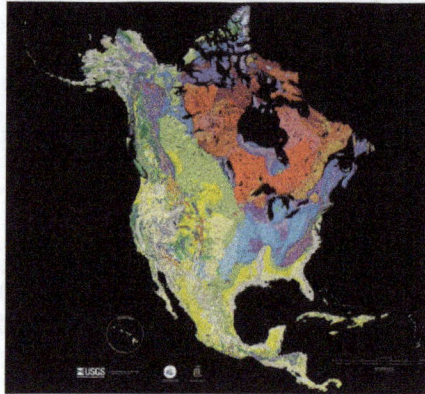

Geologic map of North America superimposed on a shaded relief map

United States

In the United States, geologic maps are usually superimposed over a topographic map (and at times over other base maps) with the addition of a color mask with letter symbols to represent the kind of geologic unit. The color mask denotes the exposure of the immediate bedrock, even if obscured by soil or other cover. Each area of color denotes a geologic unit or particular rock formation (as more information is gathered new geologic units may be defined). However, in areas where the bedrock is overlain by a significantly thick unconsolidated burden of till, terrace sediments, loess deposits, or other important feature, these are shown instead. Stratigraphic contour lines, fault lines, strike and dip symbols, are represented with various symbols as indicated by the map key. Whereas topographic maps are produced by the United States Geological Survey in conjunction with the states, geologic maps are usually produced by the individual states. There are almost no geologic map resources for some states, while a few states, such as Kentucky and Georgia, are extensively mapped geologically. Technically A map that uses colors.

United Kingdom

In the United Kingdom the term *geological map* is used. The UK and Isle of Man have been extensively mapped by the British Geological Survey (BGS) since 1835; a separate Geological Survey of Northern Ireland (drawing on BGS staff) has operated since 1947.

Two 1:625,000 scale maps cover the basic geology for the UK. More detailed sheets are available at scales of 1:250,000, 1:50,000 and 1:10,000. The 1:625,000 and 1:250,000 scales show both onshore and offshore geology (the 1:250,000 series covers the entire UK continental shelf), whilst other scales generally cover exposures on land only.

Sheets of all scales (though not for all areas) fall into two categories:

Superficial deposit maps (previously known as *solid and drift* maps) show both bedrock *and* the deposits on top of it.

Bedrock maps (previously known as *solid* maps) show the underlying rock, without superficial deposits.

The maps are superimposed over a topographic map base produced by Ordnance Survey (OS), and use symbols to represent fault lines, strike and dip or geological units, boreholes etc. Colors are used to represent different geological units. Explanatory booklets (memoirs) are produced for many sheets at the 1:50,000 scale.

Small scale thematic maps (1:1,000,000 to 1:100,000) are also produced covering geochemistry, gravity anomaly, magnetic anomaly, groundwater, etc.

Although BGS maps show the British national grid reference system and employ an OS base map, sheet boundaries are not based on the grid. The 1:50,000 sheets originate from earlier 'one inch to the mile' (1:63,360) coverage utilising the pre-grid Ordnance Survey One Inch Third Edition as the base map. Current sheets are a mixture of modern field mapping at 1:10,000 redrawn at the 1:50,000 scale and older 1:63,360 maps reproduced on a modern base map at 1:50,000. In both cases the original OS Third Edition sheet margins and numbers are retained. The 1:250,000 sheets are defined using lines of latitude and longitude, each extending 1° north-south and 2° east-west.

Singapore

The first geological map of Singapore was produced in 1974, produced by the then Public Work Department. The publication includes a locality map, 8 map sheets detailing the topography and geological units, and a sheet containing cross sections of the island.

Since 1974, for 30 years, there were many findings reported in various technical conferences on new found geology islandwide, but no new publication was produced. In 2006, Defence Science & Technology Agency, with their developments in underground space promptly started a re-publication of the Geology of Singapore, second edition.

The new edition that was published in 2009, contains a 1:75,000 geology map of the island, 6 maps (1:25,000) containing topography, street directory and geology, a sheet of cross section and a locality map.

The difference found between the 1976 Geology of Singapore report include numerous formations found in literature between 1976 and 2009. These include the Fort Canning Boulder Beds and stretches of limestone.

Geologic Modelling

Geological mapping software displaying a screenshot of a structure map generated for an 8500ft deep gas & Oil reservoir in the Erath field, Vermilion Parish, Erath, Louisiana. The left-to-right gap, near the top of the contour map indicates a Fault line. This fault line is between the blue/green contour lines and the purple/red/yellow contour lines. The thin red circular contour line in the middle of the map indicates the top of the oil reservoir. Because gas floats above oil, the thin red contour line marks the gas/oil contact zone.

Geologic modelling, Geological modelling or Geomodelling is the applied science of creating computerized representations of portions of the Earth's crust based on geophysical and geological observations made on and below the Earth surface. A Geomodel is the numerical equivalent of a three-dimensional geological map complemented by a description of physical quantities in the domain of interest. Geomodelling is related to the concept of Shared Earth Model; which is a multidisciplinary, interoperable and updatable knowledge base about the subsurface.

Geomodelling is commonly used for managing natural resources, identifying natural hazards, and quantifying geological processes, with main applications to oil and gas fields, groundwater aquifers and ore deposits. For example, in the oil and gas industry, realistic geologic models are required as input to reservoir simulator programs, which predict the behavior of the rocks under various hydrocarbon recovery scenarios. A reservoir can only be developed and produced once; therefore, making a mistake by selecting a site with poor conditions for development is tragic and wasteful. Using geological models and reservoir simulation allows reservoir engineers to identify which recovery options offer the safest and most economic, efficient, and effective development plan for a particular reservoir.

Geologic modelling is a relatively recent subdiscipline of geology which integrates structural geology, sedimentology, stratigraphy, paleoclimatology, and diagenesis;

In 2-dimensions (2D), a geologic formation or unit is represented by a polygon, which can be bounded by faults, unconformities or by its lateral extent, or crop. In geological models a geological unit is bounded by 3-dimensional (3D) triangulated or gridded surfaces. The equivalent to the mapped polygon is the fully enclosed geological unit, using a triangulated mesh. For the purpose of property or fluid modelling these volumes can be separated further into an array of cells, often referred to as voxels (volumetric elements). These 3D grids are the equivalent to 2D grids used to express properties of single surfaces.

Geomodelling generally involves the following steps:

1. Preliminary analysis of geological context of the domain of study.

2. Interpretation of available data and observations as point sets or polygonal lines (e.g. "fault sticks" corresponding to faults on a vertical seismic section).

3. Construction of a structural model describing the main rock boundaries (horizons, unconformities, intrusions, faults)

4. Definition of a three-dimensional mesh honoring the structural model to support volumetric representation of heterogeneity and solving the Partial Differential Equations which govern physical processes in the sub-surface (e.g. seismic wave propagation, fluid transport in porous media).

Geologic Modelling Components

Structural Framework

Incorporating the spatial positions of the major formation boundaries, including the effects of faulting, folding, and erosion (unconformities). The major stratigraphic divisions are further subdivided into layers of cells with differing geometries with relation to the bounding surfaces (parallel to top, parallel to base, proportional). Maximum cell dimensions are dictated by the minimum sizes of the features to be resolved (everyday example: On a digital map of a city, the location of a city park might be adequately resolved by one big green pixel, but to define the locations of the basketball court, the baseball field, and the pool, much smaller pixels - higher resolution - need to be used).

Rock Type

Each cell in the model is assigned a rock type. In a coastal clastic environment, these might be beach sand, high water energy marine upper shoreface sand, intermediate water energy marine lower shoreface sand, and deeper low energy marine silt and shale. The distribution of these rock types within the model is controlled by several methods, including map boundary polygons, rock type probability maps, or statistically emplaced based on sufficiently closely spaced well data.

Reservoir Quality

Reservoir quality parameters almost always include porosity and permeability, but may include measures of clay content, cementation factors, and other factors that affect the storage and deliverability of fluids contained in the pores of those rocks. Geostatistical techniques are most often used to populate the cells with porosity and permeability values that are appropriate for the rock type of each cell.

Fluid Saturation

Most rock is completely saturated with groundwater. Sometimes, under the right conditions, some of the pore space in the rock is occupied by other liquids or gases. In the energy industry, oil and natural gas are the fluids most commonly being modelled. The preferred methods for calculating hydrocarbon saturations in a geologic model incorporate an estimate of pore throat size, the densities of the fluids, and the height of the cell above the water contact, since these factors exert the strongest influence on capillary action, which ultimately controls fluid saturations.

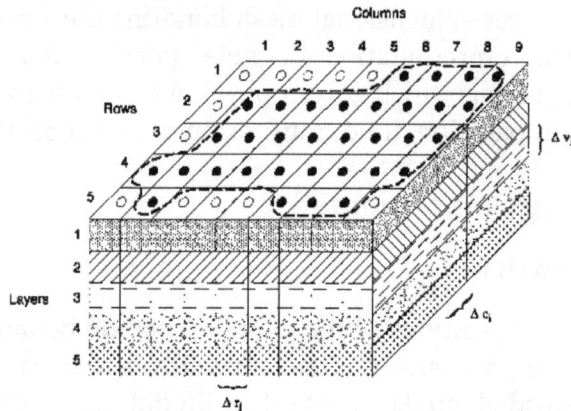

A 3D finite difference grid used in MODFLOW for simulating groundwater flow in an aquifer.

Geostatistics

An important part of geologic modelling is related to geostatistics. In order to represent the observed data, often not on regular grids, we have to use certain interpolation techniques. The most widely used technique is kriging which uses the spatial correlation among data and intends to construct the interpolation via semi-variograms. To reproduce more realistic spatial variability and help assess spatial uncertainty between data, geostatistical simulation based on variograms, training images, or parametric geological objects is often used.

Mineral Deposits

Geologists involved in mining and mineral exploration use geologic modelling to determine the geometry and placement of mineral deposits in the subsurface of the earth. Geologic models help define the volume and concentration of minerals, to which economic constraints are applied to determine the economic value of the mineralization. Mineral deposits that are deemed to be economic may be developed into a mine.

Technology

Geomodelling and CAD share a lot of common technologies. Software is usually implemented using object-oriented programming technologies in C++, Java or C# on one or multiple computer platforms. The graphical user interface generally consists of one or several 3D and 2D graphics windows to visualize spatial data, interpretations and modelling output. Such visualization is generally achieved by exploiting graphics hardware. User interaction is mostly performed through mouse and keyboard, although 3D pointing devices and immersive environments may be used in some specific cases. GIS (Geographic Information System) is also a widely used tool to manipulate geological data.

Geometric objects are represented with parametric curves and surfaces or discrete models such as polygonal meshes.

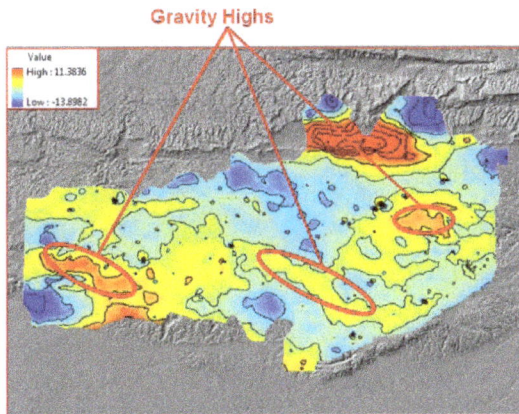

Gravity Highs

Research in Geomodelling

Problems pertainting to Geomodelling cover:

- Defining an appropriate Ontology to describe geological objects at various scales of interest,

- Integrating diverse types of observations into 3D geomodels: geological mapping data, borehole data and interpretations, seismic images and interpretations, potential field data, well test data, etc.,

- Better accounting for geological processes during model building,

- Characterizing uncertainty about the geomodels to help assess risk. Therefore, Geomodelling has a close connection to Geostatistics and Inverse problem theory,

- Applying of the recent developed Multiple Point Geostatistical Simulations (MPS) for integrating different data sources,

- Automated geometry optimization and topology conservation

History

In the 70's, geomodelling mainly consisted of automatic 2D cartographic techniques such as contouring, implemented as FORTRAN routines communicating directly with plotting hardware. The advent of workstations with 3D graphics capabilities during the 80's gave birth to a new generation of geomodelling software with graphical user interface which became mature during the 90's.

Since its inception, geomodelling has been mainly motivated and supported by oil and gas industry.

Geologic Modelling Software

Software developers have built several packages for geologic modelling purposes. Such software can display, edit, digitise and automatically calculate the parameters required by engineers, geologists and surveyors. Current software is mainly developed and commercialized by oil and gas or mining industry software vendors:

Geologic modelling and visualisation

- SGS Genesis

- IRAP RMS Suite

- Geomodeller3D

- Geosoft provides GM-SYS and VOXI 3D modelling software

- GSI3D

- Petrel

- Rockworks

- Move

Groundwater modelling

- FEFLOW

- FEHM

- MODFLOW

  - GMS

  - Visual MODFLOW

- ZOOMQ3D

Moreover, industry Consortia or companies are specifically working at improving standardization and interoperability of earth science databases and geomodelling software:

- Standardization: GeoSciML by the Commission for the Management and Application of Geoscience Information, of the International Union of Geological Sciences.

- Standardization: RESQML(tm) by Energistics

- Interoperability: OpenSpirit, by TIBCO(r)

References

- Simon Winchester, 2002, The Map that Changed the World, Harper-Collins ISBN 0-06-093180-9

- "Maclure's geological map of the United States". US Library of Congress' Map Collection. Library of Congress. Retrieved 30 October 2015.

Rock: A Major Element of Geology

Rocks are naturally formed aggregates of one or more minerals. The following text focuses on the formation of rocks and the rock cycle. The importance of rocks and their formations is a major element in the field of geology; the following text strategically encompasses and incorporates the major components and key concepts of rocks, providing a complete understanding.

Rock (Geology)

In geology, rock or stone is a naturally occurring solid aggregate of one or more minerals or mineraloids. For example, the common rock granite is a combination of the quartz, feldspar and biotite minerals. The Earth's outer solid layer, the lithosphere, is made of rock.

Balanced Rock stands in the Garden of the Gods park in Colorado Springs

Rocks have been used by mankind throughout history. From the Stone Age, rocks have been used for tools. The minerals and metals found in rocks have been essential to human civilization.

Three major groups of rocks are defined: igneous, sedimentary, and metamorphic. The scientific study of rocks is called petrology, which is an essential component of geology.

Classification

At a granular level, rocks are composed of grains of minerals, which, in turn, are homogeneous solids formed from a chemical compound that is arranged in an orderly manner. The aggregate minerals forming the rock are held together by chemical bonds. The types and abundance of minerals in a rock are determined by the manner in which the rock was formed. Many rocks contain silica (SiO_2); a compound of silicon and oxygen that forms 74.3% of the Earth's crust. This material forms crystals with other compounds in the rock. The proportion of silica in rocks and minerals is a major factor in determining their name and properties.

Rock outcrop along a mountain creek near Orosí, Costa Rica.

Rocks are geologically classified according to characteristics such as mineral and chemical composition, permeability, the texture of the constituent particles, and particle size. These physical properties are the end result of the processes that formed the rocks. Over the course of time, rocks can transform from one type into another, as described by the geological model called the rock cycle. These events produce three general classes of rock: igneous, sedimentary, and metamorphic.

The three classes of rocks are subdivided into many groups. However, there are no hard and fast boundaries between allied rocks. By increase or decrease in the proportions of their constituent minerals they pass by every gradation into one another, the distinctive structures also of one kind of rock may often be traced gradually merging into those of another. Hence the definitions adopted in establishing rock nomenclature merely correspond to more or less arbitrary selected points in a continuously graduated series.

Igneous

Igneous rock (derived from the Latin word *igneus* meaning of fire, from *ignis* meaning fire) forms through the cooling and solidification of magma or lava. This magma can be derived from partial melts of pre-existing rocks in either a planet's mantle or crust. Typically, the melting of rocks is caused by one or more of three processes: an increase in temperature, a decrease in pressure, or a change in composition.

Igneous rocks are divided into two main categories: plutonic rock and volcanic. Plutonic or intrusive rocks result when magma cools and crystallizes slowly within the Earth's crust. A common example of this type is granite. Volcanic or extrusive rocks result from magma reaching the surface either as lava or *fragmental ejecta*, forming minerals such as pumice or basalt. The chemical abundance and the rate of cooling of magma typically forms a sequence known as Bowen's reaction series. Most major igneous rocks are found along this scale.

Sample of igneous gabbro

About 64.7% of the Earth's crust by volume consists of igneous rocks; making it the most plentiful category. Of these, 66% are basalts and gabbros, 16% are granite, and 17% granodiorites and diorites. Only 0.6% are syenites and 0.3% peridotites and dunites. The oceanic crust is 99% basalt, which is an igneous rock of mafic composition. Granites and similar rocks, known as meta-granitoids, form much of the continental crust. Over 700 types of igneous rocks have been described, most of them having formed beneath the surface of Earth's crust. These have diverse properties, depending on their composition and the temperature and pressure conditions in which they were formed.

Sedimentary

Sedimentary rocks are formed at the earth's surface by the accumulation and cementation of fragments of earlier rocks, minerals, and organisms or as chemical precipitates and organic growths in water (sedimentation). This process causes clastic sediments (pieces of rock) or organic particles (detritus) to settle and accumulate, or for minerals to chemically precipitate (evaporite) from a solution. The particulate matter then undergoes compaction and cementation during at moderate temperatures and pressures (diagenesis).

Sedimentary sandstone with iron oxide bands

Before being deposited, sediments are formed by weathering or earlier rocks by erosion in a source area, and then transported to the place of deposition by water, wind, ice, mass movement or glaciers (agents of denudation). Mud rocks comprise 65% (mudstone, shale and siltstone); sandstones 20 to 25% and carbonate rocks 10 to 15% (limestone and dolostone). About 7.9% of the crust by volume is composed of sedimentary rocks, with 82% of those being shales, while the remainder consist of limestone (6%), sandstone and arkoses (12%). Sedimentary rocks often contain fossils. Sedimentary rocks form under the influence of gravity and typically are deposited in horizontal or near horizontal layers or strata and may be referred to as stratified rocks. A small fraction of sedimentary rocks deposited on steep slopes will show cross bedding where one layer stops abruptly along an interface where another layer eroded the first as it was laid atop the first.

Metamorphic

Metamorphic rocks are formed by subjecting any rock type—sedimentary rock, igneous rock or another older metamorphic rock—to different temperature and pressure conditions than those in which the original rock was formed. This process is called metamorphism; meaning to "change in form". The result is a profound change in physical properties and chemistry of the stone. The original rock, known as the protolith, transforms into other mineral types or other forms of the same minerals, by recrystallization. The temperatures and pressures required for this process are always higher than those found at the Earth's surface: temperatures greater than 150 to 200 °C and pressures of 1500 bars. Metamorphic rocks compose 27.4% of the crust by volume.

Metamorphic banded gneiss

The three major classes of metamorphic rock are based upon the formation mechanism. An intrusion of magma that heats the surrounding rock causes contact metamorphism—a temperature-dominated transformation. Pressure metamorphism occurs when sediments are buried deep under the ground; pressure is dominant and temperature plays a smaller role. This is termed burial metamorphism, and it can result in rocks such as jade. Where both heat and pressure play a role, the mechanism is termed regional metamorphism. This is typically found in mountain-building regions.

Depending on the structure, metamorphic rocks are divided into two general categories. Those that possess a texture are referred to as foliated; the remainder are termed non-foliated. The name of the rock is then determined based on the types of minerals present. Schists are foliated rocks that are primarily composed of lamellar minerals such as micas. A gneiss has visible bands of differing lightness, with a common example being the granite gneiss. Other varieties of foliated rock include slates, phyllites, and mylonite. Familiar examples of non-foliated metamorphic rocks include marble, soapstone, and serpentine. This branch contains quartzite—a metamorphosed form of sandstone—and hornfels.

Human Use

The use of rocks has had a huge impact on the cultural and technological development of the human race. Rocks have been used by humans and other hominids for at least 2.5 million years. Lithic technology marks some of the oldest and continuously used technologies. The mining of rocks for their metal ore content has been one of the most important factors of human advancement, which has progressed at different rates in different places in part because of the kind of metals available from the rocks of a region.

Ceremonial cairn of rocks, an ovoo, from Mongolia

Mi Vida uranium mine near Moab, Utah

Mining

Mining is the extraction of valuable minerals or other geological materials from the earth, from an ore body, vein or (coal) seam. This term also includes the removal of soil. Materials recovered by mining include base metals, precious metals, iron, uranium, coal, diamonds, limestone, oil shale, rock salt and potash. Mining is required to obtain any material that cannot be grown through agricultural processes, or created artificially in a laboratory or factory. Mining in a wider sense comprises extraction of any resource (e.g. petroleum, natural gas, salt or even water) from the earth.

Bay of Fires, Tasmania

Mining of rock and metals has been done since prehistoric times. Modern mining processes involve prospecting for ore bodies, analysis of the profit potential of a proposed mine, extraction of the desired materials and finally reclamation of the land to prepare it for other uses once mining ceases.

The nature of mining processes creates a potential negative impact on the environment both during the mining operations and for years after the mine has closed. This impact has led to most of the world's nations adopting regulations to manage negative effects of mining operations.

Formation of Rocks

Stone

The three main ways rocks are formed:

- Sedimentary rocks are formed through the gradual accumulation of sediments: for example, sand on a beach or mud on a river bed. As the sediments are buried they get compacted as more and more material is deposited on top. Eventually the sediments will become so dense that they would essentially form a rock. This process is known as lithification.

- Igneous rocks are rocks which have crystallized from a melt or magma. The melt is made up of various components of pre-existing rocks which have been subjected to melting either at subduction zones or within the Earth's mantle. The melt is hot and so passes upward through cooler country rock. As it moves it cools and various rock types will form through a process known as fractional crystallization. Igneous rocks can be seen at mid ocean ridges, areas of island arc volcanism or in intra-plate hotspots.

- Metamorphic rocks are rocks which once existed as igneous or sedimentary rocks but have been subjected to varying degrees of pressure and heat within the Earth's crust. The processes involved will change the composition and fabric of the rock and their original nature is often hard to distinguish. Metamorphic rocks are typically found in areas of mountain building.

Rock Synthesis

The synthetic investigation of rocks proceeds by experimental work that attempts to re-produce different rock types and to elucidate their origins and structures. In many cases no experiment is necessary. Every stage in the origin of clays, sands and gravels can be seen in process around us, but where these have been converted into coherent shales, sandstone and conglomerates, and still more where they have experienced some degree of metamorphism, there are many obscure points about their history upon which ex-periment may yet throw light. Attempts have been made to reproduce igneous rocks, by fusion of mixtures of crushed minerals or of chemicals in specially contrived furnaces. The earliest researches of this sort are those of Faujas St Fond and of de Saussure, but Sir James Hall really laid the foundations of this branch of petrology. He showed (1798) that the whinstones (diabases) of Edinburgh were fusible and if rapidly cooled yielded black vitreous masses closely resembling natural pitchstones and obsidians, if cooled more slowly they consolidated as crystalline rocks not unlike the whinstones themselves and containing olivine, augite and feldspar (the essential minerals of these rocks).

Many years later Daubrée, Delesse and others carried on similar experiments, but the first notable advance was made in 1878, when Fouqué and Lévy began their researches. They succeeded in producing such rocks as porphyrite, leucite-tephrite, basalt and dol-erite, and obtained also various structural modifications well known in igneous rocks, e.g. the porphyritic and the ophitic. Incidentally they showed that while many basic rocks (basalts, etc.) could be perfectly imitated in the laboratory, the acid rocks could not, and advanced the explanation that for the crystallization of the latter the gases never absent in natural rock magmas were indispensable mineralizing agents. It has subsequently been proved that steam, or such volatile substances as certain borates, molybdates, chlorides, fluorides, assist in the formation of orthoclase, quartz and mica (the minerals of granite). Sir James Hall also made the first contribution to the experimental study of metamorphic rocks by converting chalk into marble by heating it in a closed gun-barrel, which prevented the escape of the carbonic acid at high temperatures. In 1901 Adams and Nicholson carried this a stage further by subjecting marble to great pressures in hydraulic presses and have shown how the foliated structures, frequent in natural marble may be produced artificially.

Different Type of Rocks

Igneous Rock

Igneous rock (derived from the Latin word *ignis* meaning fire) is one of the three main rock types, the others being sedimentary and metamorphic. Igneous rock is formed through the cooling and solidification of magma or lava. Igneous rock may form with or without crystallization, either below the surface as intrusive (plutonic) rocks or on the surface as extrusive (volcanic) rocks. This magma can be derived from partial melts of existing rocks in either a planet's mantle or crust. Typically, the melting is caused by one or more of three processes: an increase in temperature, a decrease in pressure, or a change in composition.

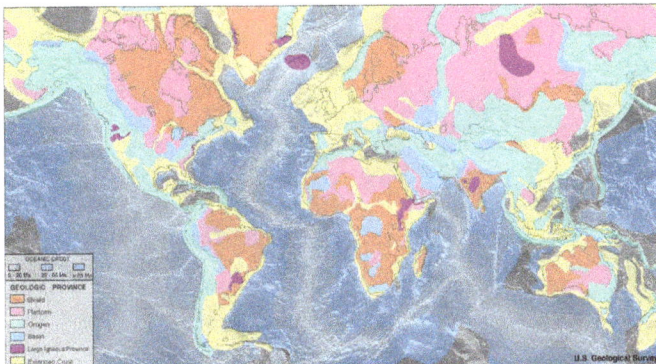

Geologic provinces of the World (USGS)

Shield	Oceanic crust:
Platform	0–20 Ma
Orogen	20–65 Ma
Basin	>65 Ma
Large igneous province	
Extended crust	

Geological Significance

Igneous and metamorphic rocks make up 90–95% of the top 16 km of the Earth's crust by volume.

Igneous rocks are geologically important because:

- their minerals and global chemistry give information about the composition of the mantle, from which some igneous rocks are extracted, and the temperature and pressure conditions that allowed this extraction, and/or of other pre-existing rock that melted;

- their absolute ages can be obtained from various forms of radiometric dating and thus can be compared to adjacent geological strata, allowing a time sequence of events;

- their features are usually characteristic of a specific tectonic environment, allowing tectonic reconstitutions;

- in some special circumstances they host important mineral deposits (ores): for example, tungsten, tin, and uranium are commonly associated with granites and diorites, whereas ores of chromium and platinum are commonly associated with gabbros.

Morphology and Setting

In terms of modes of occurrence, igneous rocks can be either intrusive (plutonic), extrusive (volcanic) or hypabyssal.

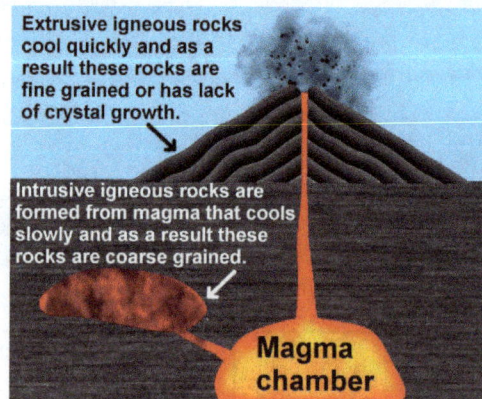

Forming of igneous rock

Intrusive

Intrusive igneous rocks are formed from magma that cools and solidifies within the crust of a planet, surrounded by pre-existing rock (called country rock); the magma cools slowly and, as a result, these rocks are coarse grained. The mineral grains in such rocks can generally be identified with the naked eye. Intrusive rocks can also be classi-fied according to the shape and size of the intrusive body and its relation to the other formations into which it intrudes. Typical intrusive formations are batholiths, stocks, laccoliths, sills and dikes. When the magma solidifies within the earth's crust, it cools slowly forming coarse textured rocks, such as granite, gabbro, or diorite.

Close-up of granite (an intrusive igneous rock) exposed in Chennai, India.

The central cores of major mountain ranges consist of intrusive igneous rocks, usually granite. When exposed by erosion, these cores (called *batholiths*) may occupy huge areas of the Earth's surface.

Coarse grained intrusive igneous rocks that form at depth within the crust are termed as *abyssal*; intrusive igneous rocks that form near the surface are termed *hypabyssal*.

Extrusive

Extrusive igneous rocks, also known as volcanic rocks, are formed at the crust's surface as a result of the partial melting of rocks within the mantle and crust. Extrusive igne-ous rocks cool and solidify quicker than intrusive igneous rocks. They are formed by the cooling of molten magma on the earth's surface. The magma, which is brought to the surface through fissures or volcanic eruptions, solidifies at a faster rate. Hence such rocks are smooth, crystalline and fine grained. Basalt is a common extrusive igneous rock and forms lava flows, lava sheets and lava plateaus. Some kinds of basalt solidify to form long polygonal columns. The Giant's Causeway in Antrim, Northern Ireland is an example.

Extrusive igneous rock is made from lava released by volcanoes

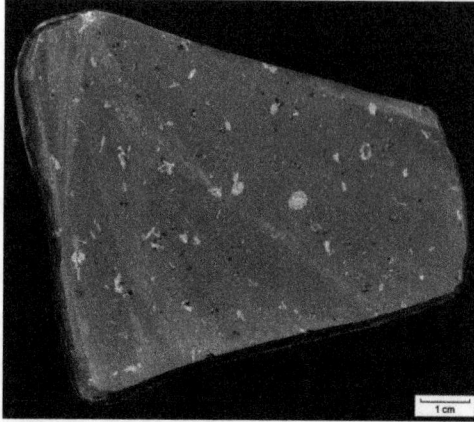

Sample of basalt (an extrusive igneous rock), found in Massachusetts

The molten rock, with or without suspended crystals and gas bubbles, is called magma. It rises because it is less dense than the rock from which it was created. When magma reaches the surface from beneath water or air, it is called lava. Eruptions of volcanoes into air are termed *subaerial*, whereas those occurring underneath the ocean are termed *submarine*. Black smokers and mid-ocean ridge basalt are examples of submarine volcanic activity.

The volume of extrusive rock erupted annually by volcanoes varies with plate tectonic setting. Extrusive rock is produced in the following proportions:

- divergent boundary: 73%

- convergent boundary (subduction zone): 15%

- hotspot: 12%.

Magma that erupts from a volcano behaves according to its viscosity, determined by temperature, composition, and crystal content. High-temperature magma, most of which is basaltic in composition, behaves in a manner similar to thick oil and, as it cools, treacle. Long, thin basalt flows with pahoehoe surfaces are common. Intermediate composition magma, such as andesite, tends to form cinder cones of intermingled ash, tuff and lava, and may have a viscosity similar to thick, cold molasses or even rubber when erupted. Felsic magma, such as rhyolite, is usually erupted at low temperature and is up to 10,000 times as viscous as basalt. Volcanoes with rhyolitic magma commonly erupt explosively, and rhyolitic lava flows are typically of limited extent and have steep margins, because the magma is so viscous.

Felsic and intermediate magmas that erupt often do so violently, with explosions driven by the release of dissolved gases—typically water vapour, but also carbon dioxide. Explosively erupted pyroclastic material is called tephra and includes tuff, agglomerate and ignimbrite. Fine volcanic ash is also erupted and forms ash tuff deposits, which can often cover vast areas.

Because lava cools and crystallizes rapidly, it is fine grained. If the cooling has been so rapid as to prevent the formation of even small crystals after extrusion, the resulting rock may be mostly glass (such as the rock obsidian). If the cooling of the lava happened more slowly, the rocks would be coarse-grained.

Because the minerals are mostly fine-grained, it is much more difficult to distinguish between the different types of extrusive igneous rocks than between different types of intrusive igneous rocks. Generally, the mineral constituents of fine-grained extrusive igneous rocks can only be determined by examination of thin sections of the rock under a microscope, so only an approximate classification can usually be made in the field.

Hypabyssal

Hypabyssal igneous rocks are formed at a depth in between the plutonic and volcanic rocks. These are formed due to cooling and resultant solidification of rising magma just beneath the earth surface. Hypabyssal rocks are less common than plutonic or volcanic rocks and often form dikes, sills, laccoliths, lopoliths, or phacoliths.

Classification

Igneous rocks are classified according to mode of occurrence, texture, mineralogy, chemical composition, and the geometry of the igneous body.

The classification of the many types of different igneous rocks can provide us with important information about the conditions under which they formed. Two important variables used for the classification of igneous rocks are particle size, which largely depends on the cooling history, and the mineral composition of the rock. Feldspars, quartz or feldspathoids, olivines, pyroxenes, amphiboles, and micas are all important minerals in the formation of almost all igneous rocks, and they are basic to the classification of these rocks. All other minerals present are regarded as nonessential in almost all igneous rocks and are called *accessory minerals*. Types of igneous rocks with other essential minerals are very rare, and these rare rocks include those with essential carbonates.

In a simplified classification, igneous rock types are separated on the basis of the type of feldspar present, the presence or absence of quartz, and in rocks with no feldspar or quartz, the type of iron or magnesium minerals present. Rocks containing quartz (silica in composition) are silica-oversaturated. Rocks with feldspathoids are silica-undersaturated, because feldspathoids cannot coexist in a stable association with quartz.

Igneous rocks that have crystals large enough to be seen by the naked eye are called phaneritic; those with crystals too small to be seen are called aphanitic. Generally speaking, phaneritic implies an intrusive origin; aphanitic an extrusive one.

An igneous rock with larger, clearly discernible crystals embedded in a finer-grained matrix is termed porphyry. Porphyritic texture develops when some of the crystals grow to considerable size before the main mass of the magma crystallizes as finer-grained, uniform material.

Igneous rocks are classified on the basis of texture and composition. Texture refers to the size, shape, and arrangement of the mineral grains or crystals of which the rock is composed.

Texture

Texture is an important criterion for the naming of volcanic rocks. The texture of volcanic rocks, including the size, shape, orientation, and distribution of mineral grains and the intergrain relationships, will determine whether the rock is termed a tuff, a pyroclastic lava or a simple lava.

Gabbro specimen showing phaneritic texture; Rock Creek Canyon, eastern Sierra Nevada, California; scale bar is 2.0 cm.

However, the texture is only a subordinate part of classifying volcanic rocks, as most often there needs to be chemical information gleaned from rocks with extremely fine-grained groundmass or from airfall tuffs, which may be formed from volcanic ash.

Textural criteria are less critical in classifying intrusive rocks where the majority of minerals will be visible to the naked eye or at least using a hand lens, magnifying glass or microscope. Plutonic rocks also tend to be less texturally varied and less prone to gaining structural fabrics. Textural terms can be used to differentiate different intrusive phases of large plutons, for instance porphyritic margins to large intrusive bodies, porphyry stocks and subvolcanic dikes (apophyses). Mineralogical classification is most often used to classify plutonic rocks. Chemical classifications are preferred to classify volcanic rocks, with phenocryst species used as a prefix, e.g. "olivine-bearing picrite" or "orthoclase-phyric rhyolite".

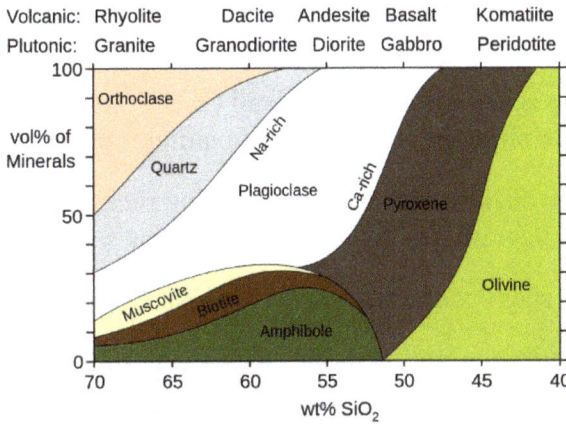

Basic classification scheme for igneous rocks on their mineralogy. If the approximate volume fractions of minerals in the rock are known, the rock name and silica content can be read off the diagram. This is not an exact method, because the classification of igneous rocks also depends on other components than silica, yet in most cases it is a good first guess.

Chemical Classification and Petrology

Igneous rocks can be classified according to chemical or mineralogical parameters.

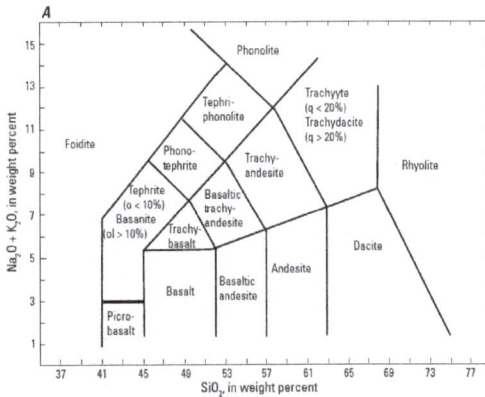

Total alkali versus silica classification scheme (TAS) as proposed in Le Maitre's 2002 Igneous Rocks - A classification and glossary of terms

Chemical: total alkali-silica content (TAS diagram) for volcanic rock classification used when modal or mineralogic data is unavailable:

- *felsic* igneous rocks containing a high silica content, greater than 63% SiO_2 (examples granite and rhyolite)

- *intermediate* igneous rocks containing between 52 – 63% SiO_2 (example andesite and dacite)

- *mafic* igneous rocks have low silica 45 – 52% and typically high iron – magnesium content (example gabbro and basalt)

- *ultramafic rock* igneous rocks with less than 45% silica. (examples picrite, komatiite and peridotite)

- *alkalic* igneous rocks with 5 – 15% alkali (K_2O + Na_2O) content or with a molar ratio of alkali to silica greater than 1:6. (examples phonolite and trachyte)

Chemical classification also extends to differentiating rocks that are chemically similar according to the TAS diagram, for instance;

- Ultrapotassic; rocks containing molar K_2O/Na_2O >3

- Peralkaline; rocks containing molar (K_2O + Na_2O)/ Al_2O_3 >1

- Peraluminous; rocks containing molar (K_2O + Na_2O)/ Al_2O_3 <1

An idealized mineralogy (the normative mineralogy) can be calculated from the chemical composition, and the calculation is useful for rocks too fine-grained or too altered for identification of minerals that crystallized from the melt. For instance, normative quartz classifies a rock as silica-oversaturated; an example is rhyolite. In an older terminology, silica oversaturated rocks were called *silicic* or *acidic* where the SiO_2 was greater than 66% and the family term *quartzolite* was applied to the most silicic. A normative feldspathoid classifies a rock as silica-undersaturated; an example is nephelinite.

History of Classification

In 1902, a group of American petrographers proposed that all existing classifications of igneous rocks should be discarded and replaced by a "quantitative" classification based on chemical analysis. They showed how vague, and often unscientific, much of the existing terminology was and argued that as the chemical composition of an igneous rock was its most fundamental characteristic, it should be elevated to prime position.

Geological occurrence, structure, mineralogical constitution—the hitherto accepted criteria for the discrimination of rock species—were relegated to the background. The completed rock analysis is first to be interpreted in terms of the rock-forming minerals which might be expected to be formed when the magma crystallizes, e.g., quartz feldspars, olivine, akermannite, Feldspathoids, magnetite, corundum, and so on, and the rocks are divided into groups strictly according to the relative proportion of these minerals to one another.

Mineralogical Classification

For volcanic rocks, mineralogy is important in classifying and naming lavas. The most important criterion is the phenocryst species, followed by the groundmass mineralogy. Often, where the groundmass is aphanitic, chemical classification must be used to properly identify a volcanic rock.

Mineralogic Contents – Felsic Versus Mafic

- *felsic* rock, highest content of silicon, with predominance of quartz, alkali feldspar and/or feldspathoids: *the felsic minerals*; these rocks (e.g., granite, rhyolite) are usually light coloured, and have low density.

- *mafic* rock, lesser content of silicon relative to felsic rocks, with predominance of mafic minerals pyroxenes, olivines and calcic plagioclase; these rocks (example, basalt, gabbro) are usually dark coloured, and have a higher density than felsic rocks.

- *ultramafic* rock, lowest content of silicon, with more than 90% of mafic minerals (e.g., dunite).

For intrusive, plutonic and usually phaneritic igneous rocks (where all minerals are visible at least via microscope), the mineralogy is used to classify the rock. This usually occurs on ternary diagrams, where the relative proportions of three minerals are used to classify the rock.

The following table is a simple subdivision of igneous rocks according to both their composition and mode of occurrence.

	Composition			
Mode of occurrence	Felsic	Intermediate	Mafic	Ultramafic
Intrusive	Granite	Diorite	Gabbro	Peridotite
Extrusive	Rhyolite	Andesite	Basalt	Komatiite

	Essential rock forming silicates			
	Felsic	Intermediate	Mafic	Ultramafic
Coarse Grained	Granite	Diorite	Gabbro	Peridotite
Medium Grained			Diabase	
Fine Grained	Rhyolite	Andesite	Basalt	Komatiite

Example of Classification

Granite is an igneous intrusive rock (crystallized at depth), with felsic composition (rich in silica and predominately quartz plus potassium-rich feldspar plus sodium-rich plagioclase) and phaneritic, subeuhedral texture (minerals are visible to the unaided eye and commonly some of them retain original crystallographic shapes).

Magma Origination

The Earth's crust averages about 35 kilometers thick under the continents, but averages only some 7–10 kilometers beneath the oceans. The continental crust is composed primarily of sedimentary rocks resting on a crystalline *basement* formed of a great variety of metamorphic and igneous rocks, including granulite and granite. Oceanic crust is composed primarily of basalt and gabbro. Both continental and oceanic crust rest on peridotite of the mantle.

Rocks may melt in response to a decrease in pressure, to a change in composition (such as an addition of water), to an increase in temperature, or to a combination of these processes.

Other mechanisms, such as melting from a meteorite impact, are less important today, but impacts during the accretion of the Earth led to extensive melting, and the outer several hundred kilometers of our early Earth was probably an ocean of magma. Impacts of large meteorites in the last few hundred million years have been proposed as one mechanism responsible for the extensive basalt magmatism of several large igneous provinces.

Decompression

Decompression melting occurs because of a decrease in pressure.

The solidus temperatures of most rocks (the temperatures below which they are completely solid) increase with increasing pressure in the absence of water. Peridotite at depth in the Earth's mantle may be hotter than its solidus temperature at some shallower level. If such rock rises during the convection of solid mantle, it will cool slightly as it expands in an adiabatic process, but the cooling is only about 0.3 °C per kilometer. Experimental studies of appropriate peridotite samples document that the solidus temperatures increase by 3 °C to 4 °C per kilometer. If the rock rises far enough, it will begin to melt. Melt droplets can coalesce into larger volumes and be intruded upwards. This process of melting from the upward movement of solid mantle is critical in the evolution of the Earth.

Decompression melting creates the ocean crust at mid-ocean ridges. It also causes volcanism in intraplate regions, such as Europe, Africa and the Pacific sea floor. There, it is variously attributed either to the rise of mantle plumes (the "Plume hypothesis") or to intraplate extension (the "Plate hypothesis").

Effects of Water and Carbon Dioxide

The change of rock composition most responsible for the creation of magma is the addition of water. Water lowers the solidus temperature of rocks at a given pressure. For example, at a depth of about 100 kilometers, peridotite begins to melt near 800 °C in the presence of excess water, but near or above about 1,500 °C in the absence of water. Water is driven out of the oceanic lithosphere in subduction zones, and it causes melting in the overlying mantle. Hydrous magmas composed of basalt and andesite are produced directly and indirectly as results of dehydration during the subduction process. Such magmas, and those derived from them, build up island arcs such as those in the

Pacific Ring of Fire. These magmas form rocks of the calc-alkaline series, an important part of the continental crust.

The addition of carbon dioxide is relatively a much less important cause of magma formation than the addition of water, but genesis of some silica-undersaturated magmas has been attributed to the dominance of carbon dioxide over water in their mantle source regions. In the presence of carbon dioxide, experiments document that the peridotite solidus temperature decreases by about 200 °C in a narrow pressure interval at pressures corresponding to a depth of about 70 km. At greater depths, carbon dioxide can have more effect: at depths to about 200 km, the temperatures of initial melting of a carbonated peridotite composition were determined to be 450 °C to 600 °C lower than for the same composition with no carbon dioxide. Magmas of rock types such as nephelinite, carbonatite, and kimberlite are among those that may be generated following an influx of carbon dioxide into mantle at depths greater than about 70 km.

Temperature Increase

Increase in temperature is the most typical mechanism for formation of magma within continental crust. Such temperature increases can occur because of the upward intrusion of magma from the mantle. Temperatures can also exceed the solidus of a crustal rock in continental crust thickened by compression at a plate boundary. The plate boundary between the Indian and Asian continental masses provides a well-studied example, as the Tibetan Plateau just north of the boundary has crust about 80 kilometers thick, roughly twice the thickness of normal continental crust. Studies of electrical resistivity deduced from magnetotelluric data have detected a layer that appears to contain silicate melt and that stretches for at least 1,000 kilometers within the middle crust along the southern margin of the Tibetan Plateau. Granite and rhyolite are types of igneous rock commonly interpreted as products of the melting of continental crust because of increases in temperature. Temperature increases also may contribute to the melting of lithosphere dragged down in a subduction zone.

Magma Evolution

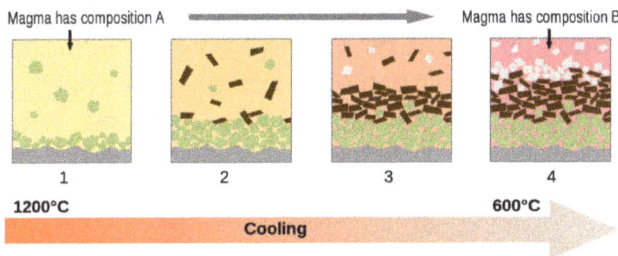

Schematic diagrams showing the principles behind fractional crystallisation in a magma. While cooling, the magma evolves in composition because different minerals crystallize from the melt. **1**: olivine crystallizes; **2**: olivine and pyroxene crystallize; **3**: pyroxene and plagioclase crystallize; **4**: plagioclase crystallizes. At the bottom of the magma reservoir, a cumulate rock forms.

Most magmas only entirely melt for small parts of their histories. More typically, they are mixes of melt and crystals, and sometimes also of gas bubbles. Melt, crystals, and bubbles usually have different densities, and so they can separate as magmas evolve.

As magma cools, minerals typically crystallize from the melt at different temperatures (fractional crystallization). As minerals crystallize, the composition of the residual melt typically changes. If crystals separate from the melt, then the residual melt will differ in composition from the parent magma. For instance, a magma of gabbroic composition can produce a residual melt of granitic composition if early formed crystals are separated from the magma. Gabbro may have a liquidus temperature near 1,200 °C, and the derivative granite-composition melt may have a liquidus temperature as low as about 700 °C. Incompatible elements are concentrated in the last residues of magma during fractional crystallization and in the first melts produced during partial melting: either process can form the magma that crystallizes to pegmatite, a rock type commonly enriched in incompatible elements. Bowen's reaction series is important for understanding the idealised sequence of fractional crystallisation of a magma.

Magma composition can be determined by processes other than partial melting and fractional crystallization. For instance, magmas commonly interact with rocks they intrude, both by melting those rocks and by reacting with them. Magmas of different compositions can mix with one another. In rare cases, melts can separate into two immiscible melts of contrasting compositions.

There are relatively few minerals that are important in the formation of common igneous rocks, because the magma from which the minerals crystallize is rich in only certain elements: silicon, oxygen, aluminium, sodium, potassium, calcium, iron, and magnesium. These are the elements that combine to form the silicate minerals, which account for over ninety percent of all igneous rocks. The chemistry of igneous rocks is expressed differently for major and minor elements and for trace elements. Contents of major and minor elements are conventionally expressed as weight percent oxides (e.g., 51% SiO_2, and 1.50% TiO_2). Abundances of trace elements are conventionally expressed as parts per million by weight (e.g., 420 ppm Ni, and 5.1 ppm Sm). The term "trace element" is typically used for elements present in most rocks at abundances less than 100 ppm or so, but some trace elements may be present in some rocks at abundances exceeding 1,000 ppm. The diversity of rock compositions has been defined by a huge mass of analytical data—over 230,000 rock analyses can be accessed on the web through a site sponsored by the U. S. National Science Foundation.

Etymology

The word "igneous" is derived from the Latin *ignis*, meaning "of fire". Volcanic rocks are named after Vulcan, the Roman name for the god of fire. Intrusive rocks are also called "plutonic" rocks, named after Pluto, the Roman god of the underworld.

Sedimentary Rock

Sedimentary rocks are types of rock that are formed by the deposition and subsequent cementation of that material at the Earth's surface and within bodies of water. Sedimentation is the collective name for processes that cause mineral and/or organic particles (detritus) to settle in place. The particles that form a sedimentary rock by accumulating are called sediment. Before being deposited, the sediment was formed by weathering and erosion from the source area, and then transported to the place of deposition by water, wind, ice, mass movement or glaciers, which are called agents of denudation. Sedimentation may also occur as minerals precipitate from water solution or shells of aquatic creatures settle out of suspension.

Middle Triassic marginal marine sequence of siltstones (reddish layers at the cliff base) and limestones (brown rocks above), Virgin Formation, southwestern Utah, USA

Sedimentary rocks on Mars, investigated by NASA's Curiosity Mars rover

Steeply dipping sedimentary rock strata along the Chalous Road in northern Iran

The sedimentary rock cover of the continents of the Earth's crust is extensive, but the total contribution of sedimentary rocks is estimated to be only 8% of the total volume of the crust. Sedimentary rocks are only a thin veneer over a crust consisting mainly of igneous and metamorphic rocks. Sedimentary rocks are deposited in layers as strata, forming a structure called bedding. The study of sedimentary rocks and rock strata provides information about the subsurface that is useful for civil engineering, for example in the construction of roads, houses, tunnels, canals or other structures. Sedimentary rocks are also important sources of natural resources like coal, fossil fuels, drinking water or ores.

The study of the sequence of sedimentary rock strata is the main source for an understanding of the Earth's history, including palaeogeography, paleoclimatology and the history of life. The scientific discipline that studies the properties and origin of sedimentary rocks is called sedimentology. Sedimentology is part of both geology and physical geography and overlaps partly with other disciplines in the Earth sciences, such as pedology, geomorphology, geochemistry and structural geology. Sedimentary rocks have also been found on Mars.

Classification Based on Origin

Sedimentary rocks can be subdivided into four groups based on the processes responsible for their formation: clastic sedimentary rocks, biochemical (biogenic) sedimentary rocks, chemical sedimentary rocks, and a fourth category for "other" sedimentary rocks formed by impacts, volcanism, and other minor processes.

Clastic Sedimentary Rocks

Clastic sedimentary rocks are composed of other rock fragments that were cemented by silicate minerals. Clastic rocks are composed largely of quartz, feldspar, rock (lithic) fragments, clay minerals, and mica; any type of mineral may be present, but they in general represent the minerals that exist locally.

Claystone deposited in Glacial Lake Missoula, Montana, United States. Note the very fine and flat bedding, common for distal lacustrine deposition.

Clastic sedimentary rocks, are subdivided according to the dominant particle size. Most geologists use the Udden-Wentworth grain size scale and divide unconsolidated sediment into three fractions: gravel (>2 mm diameter), sand (1/16 to 2 mm diameter), and mud (clay is <1/256 mm and silt is between 1/16 and 1/256 mm). The classification of clastic sedimentary rocks parallels this scheme; conglomerates and breccias are made mostly of gravel, sandstones are made mostly of sand, and mudrocks are made mostly of the finest material. This tripartite subdivision is mirrored by the broad categories of rudites, arenites, and lutites, respectively, in older literature.

The subdivision of these three broad categories is based on differences in clast shape:- conglomerates and breccias), composition (sandstones), grain size and/or texture (mudrocks).

Conglomerates and Breccias

Conglomerates are dominantly composed of rounded gravel, while breccias are composed of dominantly angular gravel.

Sandstones

Sandstone classification schemes vary widely, but most geologists have adopted the Dott scheme, which uses the relative abundance of quartz, feldspar, and lithic framework grains and the abundance of a muddy matrix between the larger grains.

Sedimentary rock with sandstone in Malta

Composition of framework grains

The relative abundance of sand-sized framework grains determines the first word in a sandstone name. Naming depends on the dominance of the three most abundant components quartz, feldspar, or the lithic fragments that originated from other rocks. All other minerals are considered accessories and not used in the naming of the rock, regardless of abundance.

- Quartz sandstones have >90% quartz grains

- Feldspathic sandstones have <90% quartz grains and more feldspar grains than lithic grains

- Lithic sandstones have <90% quartz grains and more lithic grains than feldspar grains

Abundance of muddy matrix material between sand grains

When sand-sized particles are deposited, the space between the grains either remains open or is filled with mud (silt and/or clay sized particle).

- "Clean" sandstones with open pore space (that may later be filled with matrix material) are called arenites.

- Muddy sandstones with abundant (>10%) muddy matrix are called wackes.

Six sandstone names are possible using the descriptors for grain composition (quartz-, feldspathic-, and lithic-) and the amount of matrix (wacke or arenite). For example, a quartz arenite would be composed of mostly (>90%) quartz grains and have little or no clayey matrix between the grains, a lithic wacke would have abundant lithic grains and abundant muddy matrix, etc.

Although the Dott classification scheme is widely used by sedimentologists, common names like greywacke, arkose, and quartz sandstone are still widely used by non-specialists and in popular literature.

Mudrocks

Mudrocks are sedimentary rocks composed of at least 50% silt- and clay-sized particles. These relatively fine-grained particles are commonly transported by turbulent flow in water or air, and deposited as the flow calms and the particles settle out of suspension.

Lower Antelope Canyon was carved out of the surrounding sandstone by both mechanical weathering and chemical weathering. Wind, sand, and water from flash flooding are the primary weathering agents.

Most authors presently use the term "mudrock" to refer to all rocks composed dominantly of mud. Mudrocks can be divided into siltstones, composed dominantly of silt-sized particles; mudstones with subequal mixture of silt- and clay-sized particles; and claystones, composed mostly of clay-sized particles. Most authors use "shale" as a term for a fissile mudrock (regardless of grain size) although some older literature uses the term "shale" as a synonym for mudrock.

Biochemical Sedimentary Rocks

Outcrop of Ordovician oil shale (kukersite), northern Estonia

Biochemical sedimentary rocks are created when organisms use materials dissolved in air or water to build their tissue. Examples include:

- Most types of limestone are formed from the calcareous skeletons of organisms such as corals, mollusks, and foraminifera.

- Coal, formed from plants that have removed carbon from the atmosphere and combined it with other elements to build their tissue.

- Deposits of chert formed from the accumulation of siliceous skeletons of microscopic organisms such as radiolaria and diatoms.

Chemical Sedimentary Rocks

Chemical sedimentary rock forms when mineral constituents in solution become supersaturated and inorganically precipitate. Common chemical sedimentary rocks include oolitic limestone and rocks composed of evaporite minerals, such as halite (rock salt), sylvite, barite and gypsum.

"Other" Sedimentary Rocks

This fourth miscellaneous category includes rocks formed by Pyroclastic flows, impact breccias, volcanic breccias, and other relatively uncommon processes.

Compositional Classification Schemes

Alternatively, sedimentary rocks can be subdivided into compositional groups based on their mineralogy:

- Siliciclastic sedimentary rocks, are dominantly composed of silicate minerals. The sediment that makes up these rocks was transported as bed load, suspended load, or by sediment gravity flows. Siliciclastic sedimentary rocks are subdivided into conglomerates and breccias, sandstone, and mudrocks.

- Carbonate sedimentary rocks are composed of calcite (rhombohedral $CaCO_3$), aragonite (orthorhombic $CaCO_3$), dolomite ($CaMg(CO_3)_2$), and other carbonate minerals based on the CO_{2-3} ion. Common examples include limestone and dolostone.

- Evaporite sedimentary rocks are composed of minerals formed from the evaporation of water. The most common evaporite minerals are carbonates (calcite and others based on CO_{2-3}), chlorides (halite and others built on Cl^-), and sulfates (gypsum and others built on SO_{2-4}). Evaporite rocks commonly include abundant halite (rock salt), gypsum, and anhydrite.

- Organic-rich sedimentary rocks have significant amounts of organic material, generally in excess of 3% total organic carbon. Common examples include coal, oil shale as well as source rocks for oil and natural gas.

- Siliceous sedimentary rocks are almost entirely composed of silica (SiO_2), typically as chert, opal, chalcedony or other microcrystalline forms.

- Iron-rich sedimentary rocks are composed of >15% iron; the most common forms are banded iron formations and ironstones.

- Phosphatic sedimentary rocks are composed of phosphate minerals and contain more than 6.5% phosphorus; examples include deposits of phosphate nodules, bone beds, and phosphatic mudrocks.

Deposition and Transformation

Sediment Transport and Deposition

Cross-bedding and scour in a fine sandstone; the Logan Formation
(Mississippian) of Jackson County, Ohio

Sedimentary rocks are formed when sediment is deposited out of air, ice, wind, gravity, or water flows carrying the particles in suspension. This sediment is often formed when weathering and erosion break down a rock into loose material in a source area. The material is then transported from the source area to the deposition area. The type of sediment transported depends on the geology of the hinterland (the source area of the sediment). However, some sedimentary rocks, such as evaporites, are composed of material that form at the place of deposition. The nature of a sedimentary rock, therefore, not only depends on the sediment supply, but also on the sedimentary depositional environment in which it formed.

Transformation (Diagenesis)

The term diagenesis is used to describe all the chemical, physical, and biological changes, exclusive of surface weathering, undergone by a sediment after its initial deposition. Some of those processes cause the sediment to consolidate into a compact, solid substance from the originally loose material. Young sedimentary rocks, especially those of Quaternary age (the most recent period of the geologic time scale) are often still unconsolidated. As sediment deposition builds up, the overburden (lithostatic) pressure rises, and a process known as lithification takes place.

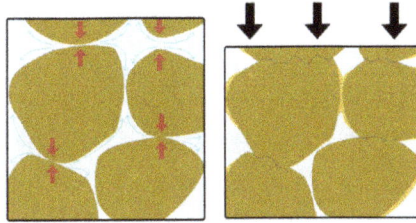

Pressure solution at work in a clastic rock. While material dissolves at places where grains are in contact, that material may recrystallize from the solution and act as cement in open pore spaces. As a result, there is a net flow of material from areas under high stress to those under low stress, producing a sedimentary rock becomes more compact and harder. Loose sand can become sandstone in this way.

Sedimentary rocks are often saturated with seawater or groundwater, in which minerals can dissolve, or from which minerals can precipitate. Precipitating minerals reduce the pore space in a rock, a process called cementation. Due to the decrease in pore space, the original connate fluids are expelled. The precipitated minerals form a cement and make the rock more compact and competent. In this way, loose clasts in a sedimentary rock can become "glued" together.

When sedimentation continues, an older rock layer becomes buried deeper as a result. The lithostatic pressure in the rock increases due to the weight of the overlying sediment. This causes compaction, a process in which grains mechanically reorganize. Compaction is, for example, an important diagenetic process in clay, which can initially consist of 60% water. During compaction, this interstitial water is pressed out of pore spaces. Compaction can also be the result of dissolution of grains by pressure solution. The dissolved material precipitates again in open pore spaces, which means there is a net flow of material into the pores. However, in some cases, a certain mineral dissolves and does not precipitate again. This process, called leaching, increases pore space in the rock.

Some biochemical processes, like the activity of bacteria, can affect minerals in a rock and are therefore seen as part of diagenesis. Fungi and plants (by their roots) and various other organisms that live beneath the surface can also influence diagenesis.

Burial of rocks due to ongoing sedimentation leads to increased pressure and temperature, which stimulates certain chemical reactions. An example is the reactions by which organic material becomes lignite or coal. When temperature and pressure increase still further, the realm of diagenesis makes way for metamorphism, the process that forms metamorphic rock.

Properties

A piece of a banded iron formation, a type of rock that consists of alternating layers with iron(III) oxide (red) and iron(II) oxide (grey). BIFs were mostly formed during the Precambrian, when the atmosphere was not yet rich in oxygen. Moories Group, Barberton Greenstone Belt, South Africa

Color

The color of a sedimentary rock is often mostly determined by iron, an element with two major oxides: iron(II) oxide and iron(III) oxide. Iron(II) oxide (FeO) only forms under low oxygen (anoxic) circumstances and gives the rock a grey or greenish colour. Iron(III) oxide (Fe_2O_3) in a richer iron environment is often found in the form of the mineral hematite and gives the rock a reddish to brownish colour. In arid continental climates rocks are in direct contact with the atmosphere, and oxidation is an important process, giving the rock a red or orange colour. Thick sequences of red sedimentary rocks formed in arid climates are called red beds. However, a red colour does not necessarily mean the rock formed in a continental environment or arid climate.

The presence of organic material can colour a rock black or grey. Organic material is formed from dead organisms, mostly plants. Normally, such material eventually decays by oxidation or bacterial activity. Under anoxic circumstances, however, organic material cannot decay and leaves a dark sediment, rich in organic material. This can, for example, occur at the bottom of deep seas and lakes. There is little water mixing in such environments, as a result oxygen from surface water is not brought down, and the deposited sediment is normally a fine dark clay. Dark rocks, rich in organic material, are therefore often shales.

Texture

The size, form and orientation of clasts (the original pieces of rock) in a sediment is called its texture. The texture is a small-scale property of a rock, but determines many of its large-scale properties, such as the density, porosity or permeability.

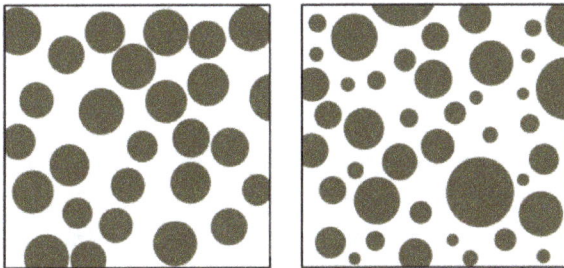

Diagram showing well-sorted (left) and poorly sorted (right) grains

The 3D orientation of the clasts is called the fabric of the rock. Between the clasts, the rock can be composed of a matrix (a cement) that consists of crystals of one or more precipitated minerals. The size and form of clasts can be used to determine the velocity and direction of current in the sedimentary environment that moved the clasts from their origin; fine, calcareous mud only settles in quiet water while gravel and larger clasts are moved only by rapidly moving water. The grain size of a rock is usually expressed with the Wentworth scale, though alternative scales are sometimes used. The grain size can be expressed as a diameter or a volume, and is always an average value – a rock is composed of clasts with different sizes. The statistical distribution of grain sizes is different for different rock types and is described in a property called the sorting of the rock. When all clasts are more or less of the same size, the rock is called 'well-sorted', and when there is a large spread in grain size, the rock is called 'poorly sorted'.

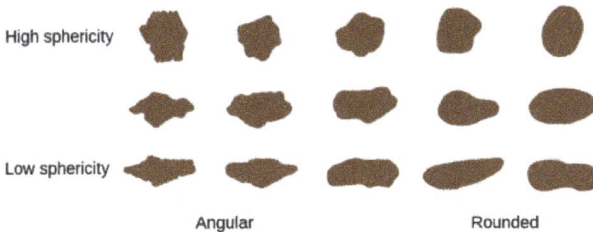

Diagram showing the rounding and sphericity of grains

The form of the clasts can reflect the origin of the rock.

Coquina, a rock composed of clasts of broken shells, can only form in energetic water. The form of a clast can be described by using four parameters:

- *Surface texture* describes the amount of small-scale relief of the surface of a grain that is too small to influence the general shape.

- *rounding* describes the general smoothness of the shape of a grain.

- 'Sphericity' describes the degree to which the grain approaches a sphere.

- 'Grain form' describes the three dimensional shape of the grain.

Chemical sedimentary rocks have a non-clastic texture, consisting entirely of crystals. To describe such a texture, only the average size of the crystals and the fabric are necessary.

Mineralogy

Most sedimentary rocks contain either quartz (especially siliciclastic rocks) or calcite (especially carbonate rocks). In contrast to igneous and metamorphic rocks, a sedimentary rock usually contains very few different major minerals. However, the origin of the minerals in a sedimentary rock is often more complex than in an igneous rock. Minerals in a sedimentary rock can have formed by precipitation during sedimentation or by diagenesis. In the second case, the mineral precipitate can have grown over an older generation of cement. A complex diagenetic history can be studied by optical mineralogy, using a petrographic microscope.

Carbonate rocks dominantly consist of carbonate minerals such as calcite, aragonite or dolomite. Both the cement and the clasts (including fossils and ooids) of a carbonate sedimentary rock can consist of carbonate minerals. The mineralogy of a clastic rock is determined by the material supplied by the source area, the manner of its transport to the place of deposition and the stability of that particular mineral. The resistance of rock forming minerals to weathering is expressed by Bowen's reaction series. In this series, quartz is the most stable, followed by feldspar, micas, and finally other less stable minerals that are only present when little weathering has occurred. The amount of weathering depends mainly on the distance to the source area, the local climate and the time it took for the sediment to be transported to the point where it is deposited. In most sedimentary rocks, mica, feldspar and less stable minerals have been reduced to clay minerals like kaolinite, illite or smectite.

Fossils

Among the three major types of rock, fossils are most commonly found in sedimentary rock. Unlike most igneous and metamorphic rocks, sedimentary rocks form at temperatures and pressures that do not destroy fossil remnants. Often these fossils may only be visible under magnification.

Fossil-rich layers in a sedimentary rock, Año Nuevo State Reserve, California

Dead organisms in nature are usually quickly removed by scavengers, bacteria, rotting and erosion, but sedimentation can contribute to exceptional circumstances where these natural processes are unable to work, causing fossilisation. The chance of fossilisation is higher when the sedimentation rate is high (so that a carcass is quickly buried), in anoxic environments (where little bacterial activity occurs) or when the organism had a particularly hard skeleton. Larger, well-preserved fossils are relatively rare.

Burrows in a turbidite, made by crustaceans, San Vincente Formation (early Eocene) of the Ainsa Basin, southern foreland of the Pyrenees

Fossils can be both the direct remains or imprints of organisms and their skeletons. Most commonly preserved are the harder parts of organisms such as bones, shells, and the woody tissue of plants. Soft tissue has a much smaller chance of being fossilized, and the preservation of soft tissue of animals older than 40 million years is very rare. Imprints of organisms made while they were still alive are called trace fossils, examples of which are burrows, footprints, etc.

As a part of a sedimentary or metamorphic rock, fossils undergo the same diagenetic processes as does the containing rock. A shell consisting of calcite can, for example, dissolve while a cement of silica then fills the cavity. In the same way, precipitating minerals can fill cavities formerly occupied by blood vessels, vascular tissue or other soft tissues. This preserves the form of the organism but changes the chemical composition, a process called permineralization. The most common minerals involved in permineralization are cements of carbonates (especially calcite), forms of amorphous silica (chalcedony, flint, chert) and pyrite. In the case of silica cements, the process is called lithification.

At high pressure and temperature, the organic material of a dead organism undergoes chemical reactions in which volatiles such as water and carbon dioxide are expelled. The fossil, in the end, consists of a thin layer of pure carbon or its mineralized form, graphite. This form of fossilisation is called carbonisation. It is particularly important for plant fossils. The same process is responsible for the formation of fossil fuels like lignite or coal.

Primary Sedimentary Structures

Cross-bedding in a fluviatile sandstone, Middle Old Red Sandstone
(Devonian) on Bressay, Shetland Islands

A flute cast, a type of sole marking, from the Book Cliffs of Utah

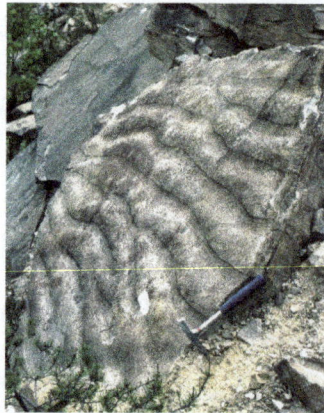

Ripple marks formed by a current in a sandstone
that was later tilted (Haßberge, Bavaria)

Structures in sedimentary rocks can be divided into 'primary' structures (formed during deposition) and 'secondary' structures (formed after deposition). Unlike textures, structures are always large-scale features that can easily be studied in the field. Sedimentary structures can indicate something about the sedimentary environment or can serve to tell which side originally faced up where tectonics have tilted or overturned sedimentary layers.

Sedimentary rocks are laid down in layers called beds or strata. A bed is defined as a layer of rock that has a uniform lithology and texture. Beds form by the deposition of layers of sediment on top of each other. The sequence of beds that characterizes sedimentary rocks is called bedding. Single beds can be a couple of centimetres to several meters thick. Finer, less pronounced layers are called laminae, and the structure it forms in a rock is called lamination. Laminae are usually less than a few centimetres thick. Though bedding and lamination are often originally horizontal in nature, this is not always the case. In some environments, beds are deposited at a (usually small) angle. Sometimes multiple sets of layers with different orientations exist in the same rock, a structure called cross-bedding. Cross-bedding forms when small-scale erosion occurs during deposition, cutting off part of the beds. Newer beds then form at an angle to older ones.

The opposite of cross-bedding is parallel lamination, where all sedimentary layering is parallel. Differences in laminations are generally caused by cyclic changes in the sediment supply, caused, for example, by seasonal changes in rainfall, temperature or biochemical activity. Laminae that represent seasonal changes (similar to tree rings) are called varves. Any sedimentary rock composed of millimeter or finer scale layers can be named with the general term *laminite*. When sedimentary rocks have no lamination at all, their structural character is called massive bedding.

Graded bedding is a structure where beds with a smaller grain size occur on top of beds with larger grains. This structure forms when fast flowing water stops flowing. Larger, heavier clasts in suspension settle first, then smaller clasts. Although graded bedding can form in many different environments, it is a characteristic of turbidity currents.

The surface of a particular bed, called the bedform, can be indicative of a particular sedimentary environment, too. Examples of bed forms include dunes and ripple marks. Sole markings, such as tool marks and flute casts, are groves dug into a sedimentary layer that are preserved. These are often elongated structures and can be used to establish the direction of the flow during deposition.

Ripple marks also form in flowing water. There are two types of ripples: symmetric and asymmetric. Environments where the current is in one direction, such as rivers, produce asymmetric ripples. The longer flank of such ripples is on the upstream side of the current. Symmetric wave ripples occur in environments where currents reverse directions, such as tidal flats.

Mudcracks are a bed form caused by the dehydration of sediment that occasionally comes above the water surface. Such structures are commonly found at tidal flats or point bars along rivers.

Secondary Sedimentary Structures

Secondary sedimentary structures are those which formed after deposition. Such structures form by chemical, physical and biological processes within the sediment. They can be indicators of circumstances after deposition. Some can be used as way up criteria.

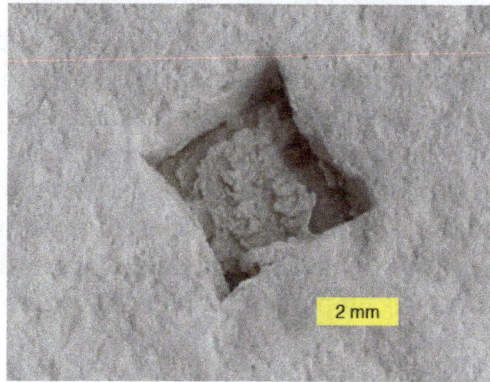

Halite crystal mold in dolomite, Paadla Formation (Silurian), Saaremaa, Estonia

Organic materials in a sediment can leave more traces than just fossils. Preserved tracks and burrows are examples of trace fossils (also called ichnofossils). Such traces are relatively rare. Most trace fossils are burrows of molluscs or arthropods. This burrowing is called bioturbation by sedimentologists. It can be a valuable indicator of the biological and ecological environment that existed after the sediment was deposited. On the other hand, the burrowing activity of organisms can destroy other (primary) structures in the sediment, making a reconstruction more difficult.

Chert concretions in chalk, Middle Lefkara Formation (upper Paleocene to middle Eocene), Cyprus

Secondary structures can also form by diagenesis or the formation of a soil (pedogenesis) when a sediment is exposed above the water level. An example of a diagenetic structure common in carbonate rocks is a stylolite. Stylolites are irregular planes where material was dissolved into the pore fluids in the rock. This can result in the precipitation of a certain chemical species producing colouring and staining of the rock, or the formation of concretions. Concretions are roughly concentric bodies with a different composition from the host rock. Their formation can be the result of localized precipitation due to small differences in composition or porosity of the host rock, such as around fossils, inside burrows or around plant roots. In carbonate based rocks such as limestone or chalk, chert or flint concretions are common, while terrestrial sandstones can have iron concretions. Calcite concretions in clay are called septarian concretions.

After deposition, physical processes can deform the sediment, producing a third class of secondary structures. Density contrasts between different sedimentary layers, such as between sand and clay, can result in flame structures or load casts, formed by inverted diapirism. While the clastic bed is still fluid, diapirism can cause a denser upper layer to sink into a lower layer. Sometimes, density contrasts can result or grow when one of the lithologies dehydrates. Clay can be easily compressed as a result of dehydration, while sand retains the same volume and becomes relatively less dense. On the other hand, when the pore fluid pressure in a sand layer surpasses a critical point, the sand can break through overlying clay layers and flow through, forming discordant bodies of sedimentary rock called sedimentary dykes. The same process can form mud volcanoes on the surface where they broke through upper layers.

Sedimentary dykes can also be formed in a cold climate where the soil is permanently frozen during a large part of the year. Frost weathering can form cracks in the soil that fill with rubble from above. Such structures can be used as climate indicators as well as way up structures.

Density contrasts can also cause small-scale faulting, even while sedimentation progresses (synchronous-sedimentary faulting). Such faulting can also occur when large masses of non-lithified sediment are deposited on a slope, such as at the front side of a delta or the continental slope. Instabilities in such sediments can result in the deposited material to slump, producing fissures and folding. The resulting structures in the rock are syn-sedimentary folds and faults, which can be difficult to distinguish from folds and faults formed by tectonic forces acting on lithified rocks.

Sedimentary Environments

The setting in which a sedimentary rock forms is called the sedimentary environment. Every environment has a characteristic combination of geologic processes and circumstances. The type of sediment that is deposited is not only dependent on the sediment that is transported to a place, but also on the environment itself.

A marine environment means that the rock was formed in a sea or ocean. Often, a distinction is made between deep and shallow marine environments. Deep marine usually refers to environments more than 200 m below the water surface. Shallow marine environments exist adjacent to coastlines and can extend to the boundaries of the continental shelf. The water movements in such environments have a generally higher energy than that in deep environments, as wave activity diminishes with depth. This means that coarser sediment particles can be transported and the deposited sediment can be coarser than in deeper environments. When the sediment is transported from the continent, an alternation of sand, clay and silt is deposited. When the continent is far away, the amount of such sediment deposited may be small, and biochemical processes dominate the type of rock that forms. Especially in warm climates, shallow marine environments far offshore mainly see deposition of carbonate rocks. The shallow, warm water is an ideal habitat for many small organisms that build carbonate skeletons. When these organisms die, their skeletons sink to the bottom, forming a thick layer of calcareous mud that may lithify into limestone. Warm shallow marine environments also are ideal environments for coral reefs, where the sediment consists mainly of the calcareous skeletons of larger organisms.

In deep marine environments, the water current working the sea bottom is small. Only fine particles can be transported to such places. Typically sediments depositing on the ocean floor are fine clay or small skeletons of micro-organisms. At 4 km depth, the solubility of carbonates increases dramatically (the depth zone where this happens is called the lysocline). Calcareous sediment that sinks below the lysocline dissolves, as a result no limestone can be formed below this depth. Skeletons of micro-organisms formed of silica (such as radiolarians) are not as soluble and still deposit. An example of a rock formed of silica skeletons is radiolarite. When the bottom of the sea has a small inclination, for example at the continental slopes, the sedimentary cover can become unstable, causing turbidity currents. Turbidity currents are sudden disturbances of the normally quite deep marine environment and can cause the geologically speaking instantaneous deposition of large amounts of sediment, such as sand and silt. The rock sequence formed by a turbidity current is called a turbidite.

The coast is an environment dominated by wave action. At a beach, dominantly denser sediment such as sand or gravel, often mingled with shell fragments, is deposited, while the silt and clay sized material is kept in mechanical suspension. Tidal flats and shoals are places that sometimes dry because of the tide. They are often cross-cut by gullies, where the current is strong and the grain size of the deposited sediment is larger. Where rivers enter the body of water, either on a sea or lake coast, deltas can form. These are large accumulations of sediment transported from the continent to places in front of the mouth of the river. Deltas are dominantly composed of clastic sediment (in contrast to chemical).

A sedimentary rock formed on land has a continental sedimentary environment. Examples of continental environments are lagoons, lakes, swamps, floodplains and alluvial fans. In the quiet water of swamps, lakes and lagoons, fine sediment is deposited, mingled with organic material from dead plants and animals. In rivers, the energy of the water is much greater and can transport heavier clastic material. Besides transport by water, sediment can in continental environments also be transported by wind or glaciers. Sediment transported by wind is called aeolian and is always very well sorted, while sediment transported by a glacier is called glacial till and is characterized by very poor sorting.

Aeolian deposits can be quite striking. The depositional environment of the Touchet Formation, located in the Northwestern United States, had intervening periods of aridity which resulted in a series of rhythmite layers. Erosional cracks were later infilled with layers of soil material, especially from aeolian processes. The infilled sections formed vertical inclusions in the horizontally deposited layers of the Touchet Formation, and thus provided evidence of the events that intervened over time among the forty-one layers that were deposited.

Sedimentary Facies

Sedimentary environments usually exist alongside each other in certain natural successions. A beach, where sand and gravel is deposited, is usually bounded by a deeper marine environment a little offshore, where finer sediments are deposited at the same time. Behind the beach, there can be dunes (where the dominant deposition is well sorted sand) or a lagoon (where fine clay and organic material is deposited). Every sedimentary environment has its own characteristic deposits. The typical rock formed in a certain environment is called its sedimentary facies. When sedimentary strata accumulate through time, the environment can shift, forming a change in facies in the subsurface at one location. On the other hand, when a rock layer with a certain age is followed laterally, the lithology (the type of rock) and facies eventually change.

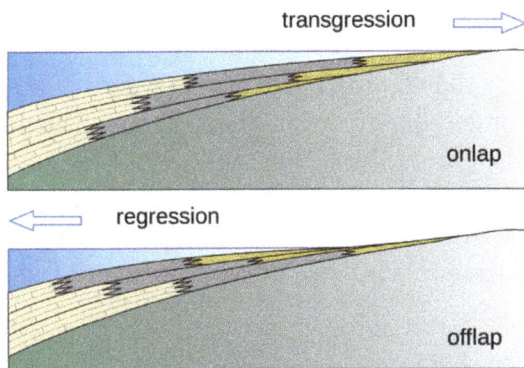

Shifting sedimentary facies in the case of transgression (above) and regression of the sea (below)

Facies can be distinguished in a number of ways: the most common are by the lithology (for example: limestone, siltstone or sandstone) or by fossil content. Coral for example only lives in warm and shallow marine environments and fossils of coral are thus typical for shallow marine facies. Facies determined by lithology are called lithofacies; facies determined by fossils are biofacies.

Sedimentary environments can shift their geographical positions through time. Coastlines can shift in the direction of the sea when the sea level drops, when the surface rises due to tectonic forces in the Earth's crust or when a river forms a large delta. In the sub-surface, such geographic shifts of sedimentary environments of the past are recorded in shifts in sedimentary facies. This means that sedimentary facies can change either parallel or perpendicular to an imaginary layer of rock with a fixed age, a phenomenon described by Walther's Law.

The situation in which coastlines move in the direction of the continent is called transgression. In the case of transgression, deeper marine facies are deposited over shallower facies, a succession called onlap. Regression is the situation in which a coastline moves in the direction of the sea. With regression, shallower facies are deposited on top of deeper facies, a situation called offlap.

The facies of all rocks of a certain age can be plotted on a map to give an overview of the palaeogeography. A sequence of maps for different ages can give an insight in the development of the regional geography.

Sedimentary Basins

Places where large-scale sedimentation takes place are called sedimentary basins. The amount of sediment that can be deposited in a basin depends on the depth of the basin, the so-called accommodation space. The depth, shape and size of a basin depend on tectonics, movements within the Earth's lithosphere. Where the lithosphere moves upward (tectonic uplift), land eventually rises above sea level, so that and erosion removes material, and the area becomes a source for new sediment. Where the lithosphere moves downward (tectonic subsidence), a basin forms and sedimentation can take place. When the lithosphere keeps subsiding, new accommodation space keeps being created.

A type of basin formed by the moving apart of two pieces of a continent is called a rift basin. Rift basins are elongated, narrow and deep basins. Due to divergent movement, the lithosphere is stretched and thinned, so that the hot asthenosphere rises and heats the overlying rift basin. Apart from continental sediments, rift basins normally also have part of their infill consisting of volcanic deposits. When the basin grows due to continued stretching of the lithosphere, the rift grows and the sea can enter, forming marine deposits.

When a piece of lithosphere that was heated and stretched cools again, its density rises, causing isostatic subsidence. If this subsidence continues long enough, the basin is called a sag basin. Examples of sag basins are the regions along passive continental margins, but sag basins can also be found in the interior of continents. In sag basins, the extra weight of the newly deposited sediments is enough to keep the subsidence going in a vicious circle. The total thickness of the sedimentary infill in a sag basins can thus exceed 10 km.

A third type of basin exists along convergent plate boundaries - places where one tectonic plate moves under another into the asthenosphere. The subducting plate bends and forms a fore-arc basin in front of the overriding plate—an elongated, deep asymmetric basin. Fore-arc basins are filled with deep marine deposits and thick sequences of turbidites. Such infill is called flysch. When the convergent movement of the two plates results in continental collision, the basin becomes shallower and develops into a foreland basin. At the same time, tectonic uplift forms a mountain belt in the overriding plate, from which large amounts of material are eroded and transported to the basin. Such erosional material of a growing mountain chain is called molasse and has either a shallow marine or a continental facies.

At the same time, the growing weight of the mountain belt can cause isostatic subsidence in the area of the overriding plate on the other side to the mountain belt. The basin type resulting from this subsidence is called a back-arc basin and is usually filled by shallow marine deposits and molasse.

Cyclic alternation of competent and less competent beds in the Blue Lias at Lyme Regis, southern England

Influence of Astronomical Cycles

In many cases facies changes and other lithological features in sequences of sedimentary rock have a cyclic nature. This cyclic nature was caused by cyclic changes in sediment supply and the sedimentary environment. Most of these cyclic changes are caused by astronomic cycles. Short astronomic cycles can be the difference between the tides or the spring tide every two weeks. On a larger time-scale, cyclic changes in climate and sea level are caused by Milankovitch cycles: cyclic changes in the orientation and/or position of the Earth's rotational axis and orbit around the Sun. There are a number of Milankovitch cycles known, lasting between 10,000 and 200,000 years.

Relatively small changes in the orientation of the Earth's axis or length of the seasons can be a major influence on the Earth's climate. An example are the ice ages of the past 2.6 million years (the Quaternary period), which are assumed to have been caused by astronomic cycles. Climate change can influence the global sea level (and thus the amount of accommodation space in sedimentary basins) and sediment supply from a certain region. Eventually, small changes in astronomic parameters can cause large changes in sedimentary environment and sedimentation.

Sedimentation Rates

The rate at which sediment is deposited differs depending on the location. A channel in a tidal flat can see the deposition of a few metres of sediment in one day, while on the deep ocean floor each year only a few millimetres of sediment accumulate. A distinction can be made between normal sedimentation and sedimentation caused by catastrophic processes. The latter category includes all kinds of sudden exceptional processes like mass movements, rock slides or flooding. Catastrophic processes can see the sudden deposition of a large amount of sediment at once. In some sedimentary environments, most of the total column of sedimentary rock was formed by catastrophic processes, even though the environment is usually a quiet place. Other sedimentary environments are dominated by normal, ongoing sedimentation.

In many cases, sedimentation occurs slowly. In a desert, for example, the wind deposits siliciclastic material (sand or silt) in some spots, or catastrophic flooding of a wadi may cause sudden deposits of large quantities of detrital material, but in most places eolian erosion dominates. The amount of sedimentary rock that forms is not only dependent on the amount of supplied material, but also on how well the material consolidates. Erosion removes most deposited sediment shortly after deposition.

Stratigraphy

That new rock layers are above older rock layers is stated in the principle of superposition. There are usually some gaps in the sequence called unconformities. These represent periods where no new sediments were laid down, or when earlier sedimentary layers were raised above sea level and eroded away.

The Permian through Jurassic stratigraphy of the Colorado Plateau area of southeastern Utah that makes up much of the famous prominent rock formations in protected areas such as Capitol Reef National Park and Canyonlands National Park. From top to bottom: Rounded tan domes of the Navajo Sandstone, layered red Kayenta Formation, cliff-forming, vertically jointed, red Wingate Sandstone, slope-forming, purplish Chinle Formation, layered, lighter-red Moenkopi Formation, and white, layered Cutler Formation sandstone. Picture from Glen Canyon National Recreation Area, Utah.

Sedimentary rocks contain important information about the history of the Earth. They contain fossils, the preserved remains of ancient plants and animals. Coal is considered a type of sedimentary rock. The composition of sediments provides us with clues as to the original rock. Differences between successive layers indicate changes to the environment over time. Sedimentary rocks can contain fossils because, unlike most igneous and metamorphic rocks, they form at temperatures and pressures that do not destroy fossil remains.

Metamorphic Rock

Metamorphic rocks arise from the transformation of existing rock types, in a process called metamorphism, which means "change in form". The original rock (protolith) is subjected to heat (temperatures greater than 150 to 200 °C) and pressure (1500 bars), causing profound physical and/or chemical change. The protolith may be a sedimentary rock, an igneous rock or another older metamorphic rock.

Quartzite, a form of metamorphic rock, from the Museum of Geology at University of Tartu collection.

Metamorphic rocks make up a large part of the Earth's crust and are classified by texture and by chemical and mineral assemblage (metamorphic facies). They may be formed simply by being deep beneath the Earth's surface, subjected to high temperatures and the great pressure of the rock layers above it. They can form from tectonic processes such as continental collisions, which cause horizontal pressure, friction and distortion. They are also formed when rock is heated up by the intrusion of hot molten rock called magma from the Earth's interior. The study of metamorphic rocks (now exposed at the Earth's surface following erosion and uplift) provides information about the temperatures and pressures that occur at great depths within the Earth's crust. Some examples of metamorphic rocks are gneiss, slate, marble, schist, and quartzite.

Metamorphic Minerals

Metamorphic minerals are those that form only at the high temperatures and pressures associated with the process of metamorphism. These minerals, known as index minerals, include sillimanite, kyanite, staurolite, andalusite, and some garnet.

Other minerals, such as olivines, pyroxenes, amphiboles, micas, feldspars, and quartz, may be found in metamorphic rocks, but are not necessarily the result of the process of metamorphism. These minerals formed during the crystallization of igneous rocks. They are stable at high temperatures and pressures and may remain chemically unchanged during the metamorphic process. However, all minerals are stable only within certain limits, and the presence of some minerals in metamorphic rocks indicates the approximate temperatures and pressures at which they formed.

The change in the particle size of the rock during the process of metamorphism is called recrystallization. For instance, the small calcite crystals in the sedimentary rock limestone and chalk change into larger crystals in the metamorphic rock marble, or in metamorphosed sandstone, recrystallization of the original quartz sand grains results in very compact quartzite, also known as metaquartzite, in which the often larger quartz crystals are interlocked. Both high temperatures and pressures contribute to recrystallization. High temperatures allow the atoms and ions in solid crystals to migrate, thus reorganizing the crystals, while high pressures cause solution of the crystals within the rock at their point of contact.

Foliation

The layering within metamorphic rocks is called *foliation* (derived from the Latin word *folia*, meaning "leaves"), and it occurs when a rock is being shortened along one axis during recrystallization. This causes the platy or elongated crystals of minerals, such as mica and chlorite, to become rotated such that their long axes are perpendicular to the orientation of shortening. This results in a banded, or foliated rock, with the bands showing the colors of the minerals that formed them.

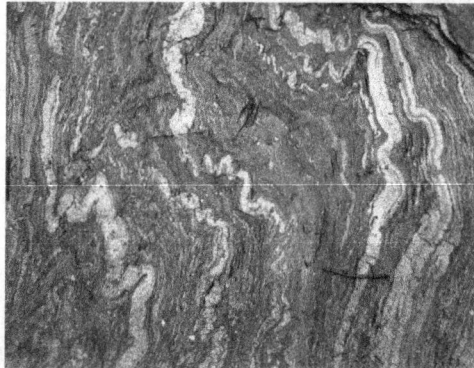

Folded foliation in a metamorphic rock from near Geirangerfjord, Norway

Textures are separated into foliated and non-foliated categories. Foliated rock is a product of differential stress that deforms the rock in one plane, sometimes creating a plane of cleavage. For example, slate is a foliated metamorphic rock, originating from shale. Non-foliated rock does not have planar patterns of strain.

Rocks that were subjected to uniform pressure from all sides, or those that lack minerals with distinctive growth habits, will not be foliated. Where a rock has been subject to differential stress, the type of foliation that develops depends on the metamorphic grade. For instance, starting with a mudstone, the following sequence develops with increasing temperature: slate is a very fine-grained, foliated metamorphic rock, characteristic of very low grade metamorphism, while phyllite is fine-grained and found in areas of low grade metamorphism, schist is medium to coarse-grained and found in areas of medium grade metamorphism, and gneiss coarse to very coarse-grained, found in areas of high-grade metamorphism. Marble is generally not foliated, which allows its use as a material for sculpture and architecture.

Another important mechanism of metamorphism is that of chemical reactions that occur between minerals without them melting. In the process atoms are exchanged between the minerals, and thus new minerals are formed. Many complex high-temperature reactions may take place, and each mineral assemblage produced provides us with a clue as to the temperatures and pressures at the time of metamorphism.

Metasomatism is the drastic change in the bulk chemical composition of a rock that often occurs during the processes of metamorphism. It is due to the introduction of chemicals from other surrounding rocks. Water may transport these chemicals rapidly over great distances. Because of the role played by water, metamorphic rocks generally contain many elements absent from the original rock, and lack some that originally were present. Still, the introduction of new chemicals is not necessary for recrystallization to occur.

Intrusive rock

Intrusive rock forms within Earth's crust from the crystallization of magma. Magma slowly pushes up from deep within the earth into any cracks or spaces it can find, sometimes pushing existing country rock out of the way, a process that can take millions of years. As the rock slowly cools into a solid, the different parts of the magma crystallize into minerals. Many mountain ranges, such as the Sierra Nevada in California, are formed mostly from large granite (or related rock) intrusions.

Devils Tower, an igneous *intrusion* exposed when the surrounding softer rock eroded away.

Intrusions are one of the two ways igneous rock can form; the other is extrusive rock, that is, a volcanic eruption or similar event. Technically speaking, an intrusion is any formation of intrusive igneous rock; rock formed from magma that cools and solidifies within the crust of the planet. In contrast, an *extrusion* consists of extrusive rock; rock formed above the surface of the crust.

Intrusions vary widely, from mountain-range-sized batholiths to thin veinlike fracture fillings of aplite or pegmatite. When exposed by erosion, these cores called batholiths may occupy huge areas of Earth's surface. Large bodies of magma that solidify underground before they reach the surface of the crust are called plutons.

Coarse-grained intrusive igneous rocks that form at depth within the earth are called abyssal while those that form near the surface are called subvolcanic or hypabyssal. Intrusive structures are often classified according to whether or not they are parallel to the bedding planes or foliation of the country rock: if the intrusion is parallel the body is concordant, otherwise it is discordant.

A well-known example of an intrusion is Devils Tower.

Structural Types

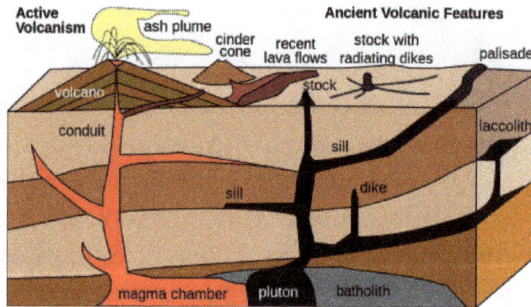

Diagram showing various types of igneous intrusion.

A *dike* intrudes into the country rock, Baranof Island, Alaska, United States.

Intrusions can be classified according to the shape and size of the intrusive body and its relation to the other formations into which it intrudes:

Batholith: a large irregular discordant intrusion

Dike: a relatively narrow tabular discordant body, often nearly vertical

Laccolith: concordant body with roughly flat base and convex top, usually with a feeder pipe below

Lopolith: concordant body with roughly flat top and a shallow convex base, may have a feeder dike or pipe below

Phacolith: a concordant lens-shaped pluton that typically occupies the crest of an anticline or trough of a syncline

Volcanic pipe or volcanic neck: tubular roughly vertical body that may have been a feeder vent for a volcano

Sill: a relatively thin tabular concordant body intruded along bedding planes

Stock: a smaller irregular discordant intrusive

Chonolith: an irregularly-shaped intrusion with a demonstrable base

Characteristics

Deep-seated intrusions are recognized from the way they have burst through the over-lying strata. Ramifying veins result from filled cracks, and the high temperature involved in this process is evident from the altered adjacent country rock. Since heat dissipates slowly and since the rock is under pressure, crystals form and no vitreous rapidly chilled matter is present. As the intrusions have had time to rest before crystallizing, they are not fluidal. Their contained gases have not been able to escape through the thick layer of strata, beneath which they were injected. Such gases form cavities, which can often be observed in these minerals. Such gases have also resulted in many important modifications in the crystallization of the rock. Because their crystals are of approximately equal size these rocks are said to be granular.

An intrusion (pink Notch Peak monzonite) inter-fingers (partly as a dike) with highly metamorphosed black-and-white-striped host rock (Cambrian carbonate rocks). Near Notch Peak, House Range, Utah.

There is typically no distinction between a first generation of large well-shaped crystals and a fine-grained ground-mass. The minerals of each have formed in a definite order, and each has had a period of crystallization that may be very distinct or may have co-incided with or overlapped the period of formation of some of the other ingredients. Earlier crystals originated at a time when most of the rock was still liquid and are more or less perfect. Later crystals are less regular in shape because they were compelled to occupy the spaces left between the already-formed crystals. The former case is said to be idiomorphic (or *automorphic*); the latter is xenomorphic. There are also many other characteristics that serve to distinguish the members of these two groups. For example, orthoclase is typically feldspar from granite, while its modifications occur in lavas of similar composition. The same distinction holds for nepheline varieties. Leucite is common in lavas but very rare in plutonic rocks. Muscovite is confined to intrusions. These differences show the influence of the physical conditions under which consolidation takes place.

Intrusive rocks formed at greater depths are called plutonic or *abyssal*. Some intrusive rocks solidified in fissures as dikes and intrusive sills at a shallow depth beneath the surface and are called subvolcanic or hypabyssal. As might be expected, they show structures intermediate between those of extrusive and plutonic rocks. They are very commonly porphyritic, vitreous, and sometimes even vesicular. In fact, many of them are petrologically indistinguishable from lavas of similar composition.

Rock Cycle

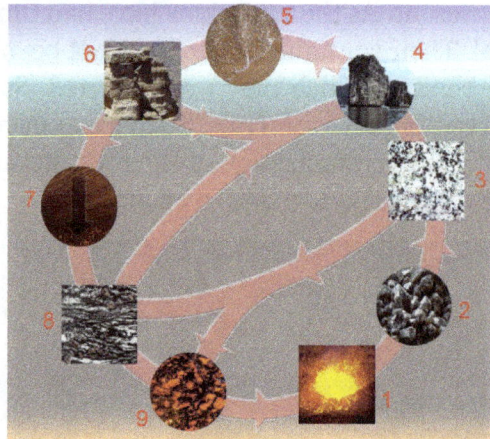

A diagram of the rock cycle. Legend: 1 = magma; 2 = crystallization (freezing of rock); 3 = igneous rocks; 4 = erosion; 5 = sedimentation; 6 = sediments & sedimentary rocks; 7 = tectonic burial and metamorphism; 8 = metamorphic rocks; 9 = melting.

The rock cycle is a basic concept in geology that describes the time-consuming transitions through geologic time among the three main rock types: sedimentary, metamorphic, and igneous. As the diagram to the right illustrates, each of the types of rocks is altered or destroyed when it is forced out of its equilibrium conditions. An igneous rock such as basalt may break down and dissolve when exposed to the atmosphere, or melt as it is subducted under a continent. Due to the driving forces of the rock cycle, plate tectonics and the water cycle, rocks do not remain in equilibrium and are forced to change as they encounter new environments. The rock cycle is an illustration that explains how the three rock types are related to each other, and how processes change from one type to another over time.

Historical Development

The original concept of the *rock cycle* is usually attributed to James Hutton, from the eighteenth century *Father of Geology*. The rock cycle was a part of Hutton's uniformitarianism and his famous quote: *no vestige of a beginning, and no prospect of an end*, applied in particular to the rock cycle and the envisioned cyclical nature of geologic processes. This concept of a repetitive non-evolutionary rock cycle remained dominant until the plate tectonics revolution of the 1960s. With the developing understanding of the driving *engine* of plate tectonics, the rock cycle changed from endlessly repetitive to a gradually evolving process. The *Wilson cycle* (a plate tectonics based rock cycle) was developed by J. Tuzo Wilson during the 1950s and 1960s. The world is made out of rocks.

The Rock Cycle

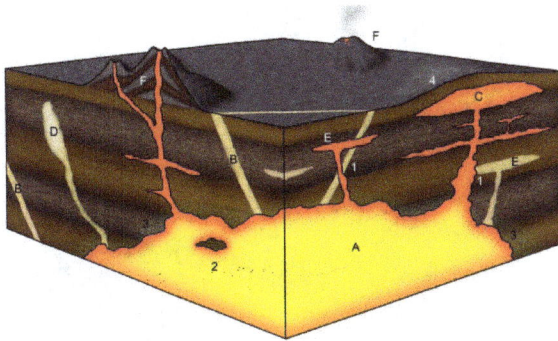

Structures of Igneous Rock. Legend: A = magma chamber (batholith); B = dyke/dike; C = laccolith; D = pegmatite; E = sill; F = stratovolcano; processes: 1 = newer intrusion cutting through older one; 2 = xenolith or roof pendant; 3 = contact metamorphism; 4 = uplift due to laccolith emplacement.

Transition to Igneous

When rocks are pushed deep under the Earth's surface, they may melt into magma. If the conditions no longer exist for the magma to stay in its liquid state, it will cool and solidify into an igneous rock. A rock that cools within the Earth is called intrusive or plutonic and will cool very slowly, producing a coarse-grained texture. As a result of volcanic activity, magma (which is called lava when it reaches Earth's surface) may cool very rapidly while being on the Earth's surface exposed to the atmosphere and are called extrusive or volcanic rocks. These rocks are fine-grained and sometimes cool so rapidly that no crystals can form and result in a natural glass, such as obsidian. Any of the three main types of rocks (igneous, sedimentary, and metamorphic rocks) can melt into magma and cool into igneous rocks.

Secondary Changes

Epigenetic change (secondary processes) may be arranged under a number of headings, each of which is typical of a group of rocks or rock-forming minerals, though usually more than one of these alterations will be found in progress in the same rock. Silicification, the replacement of the minerals by crystalline or crypto-crystalline silica, is most common in felsic rocks, such as rhyolite, but is also found in serpentine, etc. Kaolinization is the decomposition of the feldspars, which are the most common minerals in igneous rocks, into kaolin (along with quartz and other clay minerals); it is best shown by granites and syenites. Serpentinization is the alteration of olivine to serpentine (with magnetite); it is typical of peridotites, but occurs in most of the mafic rocks. In uralitization, secondary hornblende replaces augite; chloritization is the alteration of augite (biotite or hornblende) to chlorite, and is seen in many diabases, diorites and greenstones. Epidotization occurs also in rocks of this group, and consists in the development of epidote from biotite, hornblende, augite or plagioclase feldspar.

Transition to Metamorphic

This diamond is a mineral from within an igneous or metamorphic rock that formed at high temperature and pressure.

Rocks exposed to high temperatures and pressures can be changed physically or chemically to form a different rock, called metamorphic. Regional metamorphism refers to the effects on large masses of rocks over a wide area, typically associated with mountain building events within orogenic belts. These rocks commonly exhibit distinct bands of differing mineralogy and colors, called foliation. Another main type of metamorphism is caused when a body of rock comes into contact with an igneous intrusion that heats up this surrounding country rock. This *contact metamorphism* results in a rock that is altered and re-crystallized by the extreme heat of the magma and/or by the addition of fluids from the magma that add chemicals to the surrounding rock (metasomatism). Any pre-existing type of rock can be modified by the processes of metamorphism.

Transition to Sedimentary

Rocks exposed to the atmosphere are variably unstable and subject to the processes of weathering and erosion. Weathering and erosion break the original rock down into smaller fragments and carry away dissolved material. This fragmented material accumulates and is buried by additional material. While an individual grain of sand is still a member of the class of rock it was formed from, a rock made up of such grains fused together is sedimentary. Sedimentary rocks can be formed from the lithification of these buried smaller fragments (clastic sedimentary rock), the accumulation and lithification of material generated by living organisms (biogenic sedimentary rock - fossils), or lithification of chemically precipitated material from a mineral bearing solution due to evaporation (precipitate sedimentary rock). Clastic rocks can be formed from fragments broken apart from larger rocks of any type, due to processes such as erosion or from organic material, like plant remains. Biogenic and precipitate rocks form from the deposition of minerals from chemicals dissolved from all other rock types.

Forces that Drive the Rock Cycle

Plate Tectonics

In 1967, J. Tuzo Wilson published an article in Nature describing the repeated opening and closing of ocean basins, in particular focusing on the current Atlantic Ocean area. This concept, a part of the plate tectonics revolution, became known as the *Wilson cycle*. The Wilson cycle has had profound effects on the modern interpretation of the rock cycle as plate tectonics became recognized as the driving force for the rock cycle.

Spreading Ridges

At the mid-ocean divergent boundaries *new* magma is produced by mantle upwelling and a shallow *melting zone*. This *juvenile* basaltic magma is an early phase of the igneous portion of the cycle. As the tectonic plates on either side of the ridge move apart the new rock is carried away from the ridge, the interaction of heated circulating seawater through fractures starts the retrograde metamorphism of the new rock.

Subduction Zones

The new basaltic oceanic crust eventually meets a subduction zone as it moves away from the spreading ridge. As this crust is pulled back into the mantle, the increasing pressure and temperature conditions cause a restructuring of the mineralogy of the rock, this metamorphism alters the rock to form eclogite. As the slab of basaltic crust and some included sediments are dragged deeper, water and other more volatile materials are driven off and rise into the overlying wedge of rock above the subduction zone which is at a lower pressure. The lower pressure, high temperature, and now volatile rich material in this wedge melts and the resulting buoyant magma rises through the overlying rock to produce island arc or continental margin volcanism. This volcanism includes more silicic lavas the further from the edge of the island arc or continental margin, indicating a deeper source and a more differentiated magma.

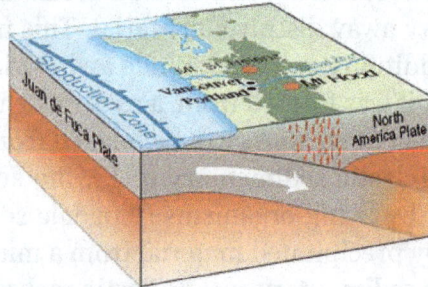

The Juan de Fuca plate sinks below the North America plate at the Cascadia subduction zone.

At times some of the metamorphosed downgoing slab may be thrust up or obducted onto the continental margin. These blocks of mantle peridotite and the metamorphic eclogites are exposed as ophiolite complexes.

The newly erupted volcanic material is subject to rapid erosion depending on the climate conditions. These sediments accumulate within the basins on either side of an island arc. As the sediments become more deeply buried lithification begins and sedimentary rock results.

Continental Collision

On the closing phase of the classic Wilson cycle, two continental or smaller terranes meet at a convergent zone. As the two masses of continental crust meet, neither can be subducted as they are both *low density* silicic rock. As the two masses meet, tremendous compressional forces distort and modify the rocks involved. The result is regional metamorphism within the interior of the ensuing orogeny or mountain building event. As the two masses are compressed, folded and faulted into a mountain range by the continental collision the whole suite of pre-existing igneous, volcanic, sedimentary and earlier metamorphic rock units are subjected to this new metamorphic event.

Accelerated Erosion

The high mountain ranges produced by continental collisions are immediately subjected to the forces of erosion. Erosion wears down the mountains and massive piles of sediment are developed in adjacent ocean margins, shallow seas, and as continental deposits. As these sediment piles are buried deeper they become lithified into sedimentary rock. The metamorphic, igneous, and sedimentary rocks of the mountains become the new piles of sediments in the adjoining basins and eventually become sedimentary rock.

An Evolving Process

The plate tectonics rock cycle is an evolutionary process. Magma generation, both in the spreading ridge environment and within the wedge above a subduction zone, favors the eruption of the more silicic and volatile rich fraction of the crustal or upper mantle material. This lower density material tends to stay within the crust and not be subducted back into the mantle. The magmatic aspects of plate tectonics tends to gradual segregation within or between the mantle and crust. As magma forms, the initial melt is composed of the more silicic phases that have a lower melting point. This leads to partial melting and further segregation of the lithosphere. In addition the silicic continental crust is relatively buoyant and is not normally subducted back into the mantle. So over time the continental masses grow larger and larger.

The Role of Water

The presence of abundant water on Earth is of great importance for the rock cycle. Most obvious perhaps are the water driven processes of weathering and erosion. Water in the form of precipitation and acidic soil water and groundwater is quite effective at dissolving minerals and rocks, especially those igneous and metamorphic rocks and marine sedimentary rocks that are unstable under near surface and atmospheric conditions. The water carries away the ions dissolved in solution and the broken down fragments that are the products of weathering. Running water carries vast amounts of sediment in rivers back to the ocean and inland basins. The accumulated and buried sediments are converted back into rock.

A less obvious role of water is in the metamorphism processes that occur in fresh seafloor volcanic rocks as seawater, sometimes heated, flows through the fractures and crevices in the rock. All of these processes, illustrated by serpentinization, are an important part of the destruction of volcanic rock.

The role of water and other volatiles in the melting of existing crustal rock in the wedge above a subduction zone is a most important part of the cycle. Along with water, the presence of carbon dioxide and other carbon compounds from abundant marine limestone within the sediments atop the down going slab is another source of melt inducing volatiles. This involves the carbon cycle as a part of the overall rock cycle.

References

- Wilson, James Robert (1995), A collector's guide to rock, mineral & fossil localities of Utah, Utah Geological Survey, pp. 1–22, ISBN 1557913366.

- Bucher, Kurt; Grapes, Rodney (2011), Petrogenesis of Metamorphic Rocks, Springer, pp. 23–24, ISBN 3540741682.

- Botin, J.A., ed. (2009). Sustainable Management of Mining Operations. Denver, CO, USA: Society for Mining, Metallurgy, and Exploration. ISBN 9780873352673.

- Wilson, Arthur (1996). The Living Rock: The Story of Metals Since Earliest Times and Their Impact on Developing Civilization. Cambridge, England: Woodhead Publishing. ISBN 1855733013.

- Prothero, Donald R.; Schwab, Fred (2004). Sedimentary geology : an introduction to sedimentary rocks and stratigraphy (2nd ed.). New York: Freeman. p. 12. ISBN 978-0-7167-3905-0.

- Geoff C. Brown; C. J. Hawkesworth; R. C. L. Wilson (1992). Understanding the Earth (2nd ed.). Cambridge University Press. p. 93. ISBN 0-521-42740-1.

- Blatt, Harvey & Robert J. Tracy (1996). Petrology; Igneous, Sedimentary, and Metamorphic, 2nd Ed. W. H. Freeman. ISBN 0-7167-2438-3.

- Terrascope. "Environmental Risks of Mining". The Future of strategic Natural Resources. Cambridge, MA, USA: Massachusetts Institute of Technology. Retrieved 10 September 2014.

- Roberts, Dar. "Rocks and classifications". Department of Geography, University of California, Santa Barbara. Retrieved 11 November 2012.

History of Geology

The history of geology concerns itself with the development that has taken place in the study of geology over a period of years. Geology studies the origin, history and structure of the earth. The section has been carefully written to provide an easy understanding of the history of geology.

History of Geology

The history of geology is concerned with the development of the natural science of geology. Geology is the scientific study of the origin, history, and structure of the Earth.

Scotsman James Hutton is considered to be the father of modern geology

Antiquity

Some of the first geological thoughts were about the origin of the Earth. Ancient Greece developed some primary geological concepts concerning the origin of the Earth. Additionally, in the 4th century BC Aristotle made critical observations of the slow rate of geological change. He observed the composition of the land and formulated a theory where the Earth changes at a slow rate and that these changes cannot be observed during one person's lifetime. Aristotle developed one of the first evidentially based concepts connected to the geological realm regarding the rate at which the Earth physically changes.

A mosquito and a fly in this Baltic amber necklace are between 40 and 60 million years old

The slightly misshapen octahedral shape of this rough diamond crystal in matrix is typical of the mineral. Its lustrous faces also indicate that this crystal is from a primary deposit.

However, it was his successor at the Lyceum, the philosopher Theophrastus, who made the greatest progress in antiquity in his work *On Stones*. He described many minerals and ores both from local mines such as those at Laurium near Athens, and further afield. He also quite naturally discussed types of marble and building materials like limestones, and attempted a primitive classification of the properties of minerals by their properties such as hardness.

Much later in the Roman period, Pliny the Elder produced a very extensive discussion of many more minerals and metals then widely used for practical ends. He was among the first to correctly identify the origin of amber as a fossilized resin from trees by the observation of insects trapped within some pieces. He also laid the basis of crystallography by recognising the octahedral habit of diamond.

Middle Ages

Abu al-Rayhan al-Biruni (AD 973-1048) was one of the earliest Muslim geologists, whose works included the earliest writings on the geology of India, hypothesizing that the Indian subcontinent was once a sea:

Ibn Sina (Avicenna, 981-1037), a Persian polymath, made significant contributions to geology and the natural sciences (which he called *Attabieyat*) along with other natural philosophers such as Ikhwan AI-Safa and many others. Ibn Sina wrote an encyclopaedic work entitled *"Kitab al-Shifa"* (the Book of Cure, Healing or Remedy from ignorance), in which Part 2, Section 5, contains his commentary on Aristotle's Mineralogy and Meteorology, in six chapters: Formation of mountains, The advantages of mountains in the formation of clouds; Sources of water; Origin of earthquakes; Formation of minerals; The diversity of earth's terrain.

In medieval China, one of the most intriguing naturalists was Shen Kuo (1031-1095), a polymath personality who dabbled in many fields of study in his age. In terms of geology, Shen Kuo is one of the first naturalists to have formulated a theory of geomorphology. This was based on his observations of sedimentary uplift, soil erosion, deposition of silt, and marine fossils found in the Taihang Mountains, located hundreds of miles from the Pacific Ocean. He also formulated a theory of gradual climate change, after his observation of ancient petrified bamboos found in a preserved state underground near Yanzhou (modern Yan'an), in the dry northern climate of Shaanxi province. He formulated a hypothesis for the process of land formation: based on his observation of fossil shells in a geological stratum in a mountain hundreds of miles from the ocean, he inferred that the land was formed by erosion of the mountains and by deposition of silt.

17th Century

A portrait of Whiston with a diagram demonstrating his theories of cometary catastrophism best described in *A New Theory of the Earth*

It was not until the 17th century that geology made great strides in its development. At this time, geology became its own entity in the world of natural science. It was discovered by the Christian world that different translations of the Bible contained different versions of the biblical text. The one entity that remained consistent through all of the interpretations was that the Deluge had formed the world's geology and geography. To prove the Bible's authenticity, individuals felt the need to demonstrate with scientific evidence that the Great Flood had in fact occurred. With this enhanced desire for data came an increase in observations of the Earth's composition, which in turn led to the discovery of fossils. Although theories that resulted from the heightened interest in the

Earth's composition were often manipulated to support the concept of the Deluge, a genuine outcome was a greater interest in the makeup of the Earth. Due to the strength of Christian beliefs during the 17th century, the theory of the origin of the Earth that was most widely accepted was *A New Theory of the Earth* published in 1696, by William Whiston. Whiston used Christian reasoning to "prove" that the Great Flood had occurred and that the flood had formed the rock strata of the Earth.

During the 17th century the heated debate between religion and science over the Earth's origin further propelled interest in the Earth and brought about more systematic identification techniques of the Earth's strata. The Earth's strata can be defined as horizontal layers of rock having approximately the same composition throughout. An important pioneer in the science was Nicolas Steno. Steno was trained in the classical texts on science; however, by 1659 he seriously questioned accepted knowledge of the natural world. Importantly, he questioned the idea that fossils grew in the ground, as well as common explanations of rock formation. His investigations and his subsequent conclusions on these topics have led scholars to consider him one of the founders of modern stratigraphy and geology.

18th Century

From this increased interest in the nature of the Earth and its origin, came a heightened attention to minerals and other components of the Earth's crust. Moreover, the increasing economic importance of mining in Europe during the mid to late 18th century made the possession of accurate knowledge about ores and their natural distribution vital. Scholars began to study the makeup of the Earth in a systematic manner, with detailed comparisons and descriptions not only of the land itself, but of the semi-precious metals it contained, which had great commercial value. For example, in 1774 Abraham Gottlob Werner published the book *Von den äusserlichen Kennzeichen der Fossilien (On the External Characters of Minerals),* which brought him widespread recognition because he presented a detailed system for identifying specific minerals based on external characteristics. The more efficiently productive land for mining could be identified and the semi-precious metals could be found, the more money could be made. This drive for economic gain propelled geology into the limelight and made it a popular subject to pursue. With an increased number of people studying it, came more detailed observations and more information about the Earth.

Also during the eighteenth century, aspects of the history of the Earth—namely the divergences between the accepted religious concept and factual evidence—once again became a popular topic for discussion in society. In 1749 the French naturalist Georges-Louis Leclerc, Comte de Buffon published his *Histoire Naturelle,* in which he attacked the popular Biblical accounts given by Whiston and other ecclesiastical theorists of the history of Earth. From experimentation with cooling globes, he found that the age of the Earth was not only 4,000 or 5,500 years as inferred from the Bible, but rather 75,000 years. Another individual who described the history of the Earth with reference to neither God nor the Bible was the philosopher Immanuel Kant, who published his *Universal Natural History and Theory of the Heavens (Allgemeine Naturgeschichte und Theorie des Himmels)* in 1755. From the works of these respected men, as well as others, it became acceptable by the mid eighteenth century to question the age of the Earth. This questioning represented a turning point in the study of the Earth. It was now possible to study the history of the Earth from a scientific perspective without religious preconceptions.

With the application of scientific methods to the investigation of the Earth's history, the study of geology could become a distinct field of science. To begin with, the terminology and definition of what constituted geological study had to be worked out. The term "geology" was first used technically in publications by two Genevan naturalists, Jean-André Deluc and Horace-Bénédict de Saussure, though "geology" was not well received as a term until it was taken up in the very influential compendium, the *Encyclopédie*, published beginning in 1751 by Denis Diderot. Once the term was established to denote the study of the Earth and its history, geology slowly became more generally recognized as a distinct science that could be taught as a field of study at educational institutions. In 1741 the best-known institution in the field of natural history, the National Museum of Natural History in France, created the first teaching position designated specifically for geology. This was an important step in further promoting knowledge of geology as a science and in recognizing the value of widely disseminating such knowledge.

By the 1770s chemistry was starting to play a pivotal role in the theoretical foundation of geology and two opposite theories with committed followers emerged. These contrasting theories offered differing explanations of how the rock layers of the Earth's surface had formed. One suggested that a liquid inundation, perhaps like the biblical deluge, had created all geological strata. The theory extended chemical theories that had been developing since the seventeenth century and was promoted by Scotland's John Walker, Sweden's Johan Gottschalk Wallerius and Germany's Abraham Werner. Of these names, Werner's views become internationally influential around 1800. He argued that the Earth's layers, including basalt and granite, had formed as a precipitate from an ocean that covered the entire Earth. Werner's system was influential and those who accepted his theory were known as Diluvianists or Neptunists. The Neptunist thesis was the most popular during the late eighteenth century, especially for those who were chemically trained. However, another thesis slowly gained currency from the 1780s forward. Instead of water, some mid eighteenth-century naturalists such as Buffon had suggested that strata had been formed through heat (or fire). The thesis was modified and expanded by the Scottish naturalist James Hutton during the 1780s. He argued against the theory of Neptunism, proposing instead the theory of based on heat. Those who followed this thesis during the early nineteenth century referred to this view as Plutonism: the formation of the Earth through the gradual solidification of a molten mass at a slow rate by the same processes that had occurred throughout history and continued in the present day. This led him to the conclusion that the Earth was immeasurably old and could not possibly be explained within the limits of the chronology inferred from the Bible. Plutonists believed that volcanic processes were the chief agent in rock formation, not water from a Great Flood.

19th Century

Bust of William Smith, in the Oxford University Museum of Natural History.

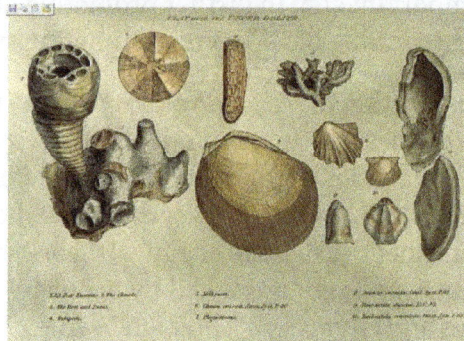

Engraving from William Smith's 1815 monograph on identifying strata by fossils

In the early 19th century the mining industry and Industrial Revolution stimulated the rapid development of the stratigraphic column - "the sequence of rock formations arranged according to their order of formation in time." In England. the mining surveyor William Smith, starting in the 1790s, found empirically that fossils were a highly effective means of distinguishing between otherwise similar formations of the landscape as he travelled the country working on the canal system and produced the first geological map of Britain. At about the same time, the French comparative anatomist Georges Cuvier assisted by his colleague Alexandre Brogniart at the École des Mines de Paris realized that the relative ages of fossils could be determined from a geological standpoint; in terms of what layer of rock the fossils are located and the distance these layers of rock are from the surface of the Earth. Through the synthesis of their findings, Brogniart and Cuvier realized that different strata could be identified by fossil contents and thus each stratum could be assigned to a unique position in a sequence. After the publication of Cuvier and Brongniart's book, "Description Geologiques des Environs de Paris" in 1811, which outlined the concept, stratigraphy became very popular amongst geologists; many hoped to apply this concept to all the rocks of the Earth. During this century various geologists further refined and completed the stratigraphic column. For instance, in 1833 while Adam Sedgwick was mapping rocks that he had established were from the Cambrian Period, Charles Lyell was elsewhere suggesting a subdivision of the Tertiary Period; whilst Roderick Murchison, mapping into Wales from a different direction, was assigning the upper parts of Sedgewick's *Cambrian* to the lower parts of his own Silurian Period. The stratigraphic column was significant because it supplied a method to assign a relative age of these rocks by slotting them into different positions in their stratigraphical sequence. This created a global approach to dating the age of the Earth and allowed for further correlations to be drawn from similarities found in the makeup of the Earth's crust in various countries.

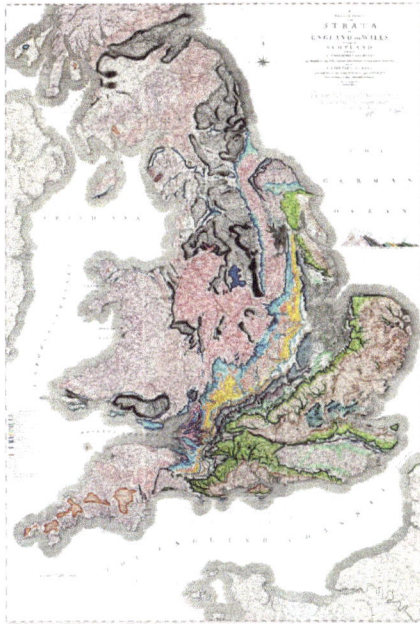

Geological map of Great Britain by William Smith, published 1815.

In early nineteenth-century Britain, catastrophism was adapted with the aim of reconciling geological science with religious traditions of the biblical Great Flood. In the early 1820s English geologists including William Buckland and Adam Sedgwick interpreted "diluvial" deposits as the outcome of Noah's flood, but by the end of the decade they revised their opinions in favour of local inundations. Charles Lyell challenged catastrophism with the publication in 1830 of the first volume of his book *Principles of Geology* which presented a variety of geological evidence from England, France, Italy and Spain to prove Hutton's ideas of gradualism correct. He argued that most geological change had been very gradual in human history. Lyell provided evidence for Uniformitarianism; a geological doctrine that processes occur at the same rates in the present as they did in the past and account for all of the Earth's geological features. Lyell's works were popular and widely read, the concept of Uniformitarianism had taken a strong hold in geological society.

During the same time that the stratigraphic column was being completed, imperialism drove several countries in the early to mid 19th century to explore distant lands to expand their empires. This gave naturalists the opportunity to collect data on these voyages. In 1831 Captain Robert FitzRoy, given charge of the coastal survey expedition of HMS *Beagle*, sought a suitable naturalist to examine the land and give geological advice. This fell to Charles Darwin, who had just completed his BA degree and had accompanied Sedgwick on a two-week Welsh mapping expedition after taking his Spring course on geology. Fitzroy gave Darwin Lyell's *Principles of Geology*, and Darwin became Lyell's first disciple, inventively theorising on uniformitarian principles about the geological processes he saw, and challenging some of Lyell's ideas. He speculated about the Earth expanding to explain uplift, then on the basis of the idea that ocean areas sank as land was uplifted, theorised that coral atolls grew from fringing coral reefs round sinking volcanic islands. This idea was confirmed when the *Beagle* surveyed the Cocos (Keeling) Islands, and in 1842 he published his theory on *The Structure and Distribution of Coral Reefs*. Darwin's discovery of giant fossils helped to establish his reputation as a geologist, and his theorising about the causes of their extinction led to his theory of evolution by natural selection published in *On the Origin of Species* in 1859.

Economic motivations for the practical use of geological data caused governments to support geological research. During the 19th century the governments of several countries including Canada, Australia, Great Britain and the United States funded geological surveying that would produce geological maps of vast areas of the countries. Geological surveying provides the location of useful minerals and such information could be used to benefit the country's mining industry. With the government funding of geological research, more individuals could study geology with better technology and techniques, leading to the expansion of the field of geology.

In the 19th century, scientific realms established the age of the Earth in terms of millions of years. By the early 20th century the Earth's estimated age was 2 billion years. Radiometric dating determined the age of minerals and rocks, which provided necessary data to help determine the Earth's age. With this new discovery based on verifiable scientific data and the possible age of the Earth extending billions of years, the dates of the geological time scale could now be refined. Theories that did not comply with the scientific evidence that established the age of the Earth could no longer be accepted.

20th Century

The determined age of the Earth as 2 billion years opened doors for theories of continental movement during this vast amount of time. In 1912 Alfred Wegener proposed the theory of Continental Drift. This theory suggests that the continents were joined together at a certain time in the past and formed a single landmass known as Pangaea; thereafter they drifted like rafts over the ocean floor, finally reaching their present position. The shapes of continents and matching coastline geology between some continents indicated they were once attached together as Pangea. Additionally, the theory of continental drift offered a possible explanation as to the formation of mountains. From this, different theories developed as to how mountains were built. Unfortunately, Wegener provided no convincing mechanism for this drift, and his ideas were not accepted during his lifetime.

Alfred Wegener, around 1925

Research from 1947 found new evidence about the ocean floor, and in 1960 Bruce C. Heezen published the concept of mid-ocean ridges. Soon after this, Robert S. Dietz and Harry H. Hess proposed that the oceanic crust forms as the seafloor spreads apart along mid-ocean ridges in seafloor spreading. This led directly to the theory of Plate Tectonics that was well supported and accepted by almost all geologists by the end of the decade, and provided a mechanism explaining the apparent drift which Wegener had proposed. Geophysical evidence suggested lateral motion of continents and that oceanic crust is younger than continental crust. This geophysical evidence also spurred the hypothesis of paleomagnetism, the record of the orientation of the Earth's magnetic field recorded in magnetic minerals. British geophysicist S. K. Runcorn suggested the concept of paleomagnetism from his finding that the continents had moved relative to the Earth's magnetic poles.

Modern Geology

By applying sound stratigraphic principles to the distribution of craters on the Moon, it can be argued that almost overnight, Gene Shoemaker took the study of the Moon away from Lunar astronomers and gave it to Lunar geologists.

In recent years, geology has continued its tradition as the study of the character and origin of the Earth, its surface features and internal structure. What changed in the later 20th century is the perspective of geological study. Geology was now studied using a more integrative approach, considering the Earth in a broader context encompassing the atmosphere, biosphere and hydrosphere. Satellites located in space that take wide scope photographs of the Earth provide such a perspective. In 1972, The Landsat Program, a series of satellite missions jointly managed by NASA and the U.S. Geological Survey, began supplying satellite images that can be geologically analyzed. These images can be used to map major geological units, recognize and correlate rock types for vast regions and track the movements of Plate Tectonics. A few applications of this data include the ability to produce geologically detailed maps, locate sources of natural energy and predict possible natural disasters caused by plate shifts.

Geological History of Earth

The geological history of Earth follows the major events in Earth's past based on the geologic time scale, a system of chronological measurement based on the study of the planet's rock layers (stratigraphy). Earth formed about 4.54 billion years ago by accretion from the solar nebula, a disk-shaped mass of dust and gas left over from the formation of the Sun, which also created the rest of the Solar System.

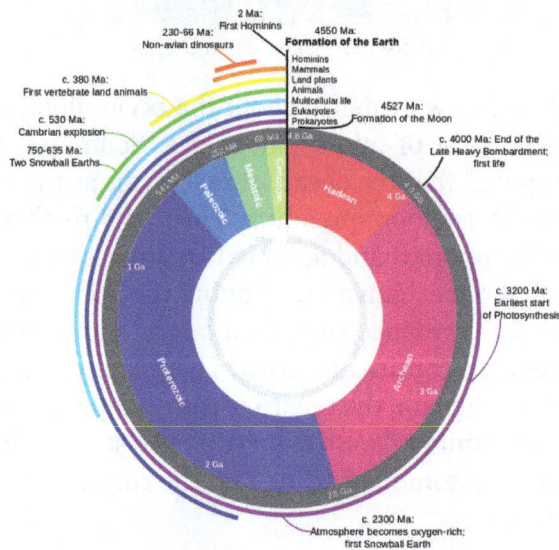

Geologic time represented in a diagram called a geological clock, showing the relative lengths of the eons of Earth's history and noting major events

Earth was initially molten due to extreme volcanism and frequent collisions with other bodies. Eventually, the outer layer of the planet cooled to form a solid crust when water began accumulating in the atmosphere. The Moon formed soon afterwards, possibly as a result of the impact of a planetoid with the Earth. Outgassing and volcanic activity produced the primordial atmosphere. Condensing water vapor, augmented by ice delivered from comets, produced the oceans.

As the surface continually reshaped itself over hundreds of millions of years, continents formed and broke apart. They migrated across the surface, occasionally combining to form a supercontinent. Roughly 750 million years ago, the earliest-known supercontinent Rodinia, began to break apart. The continents later recombined to form Pannotia, 600 to 540 million years ago, then finally Pangaea, which broke apart 200 million years ago.

The present pattern of ice ages began about 40 million years ago, then intensified at the end of the Pliocene. The polar regions have since undergone repeated cycles of glaciation and thaw, repeating every 40,000–100,000 years. The last glacial period of the current ice age ended about 10,000 years ago.

Precambrian

The Precambrian includes approximately 90% of geologic time. It extends from 4.6 billion years ago to the beginning of the Cambrian Period (about 541 Ma). It includes three eons, the Hadean, Archean, and Proterozoic.

Hadean Eon

During Hadean time (4.6–4 Ga), the Solar System was forming, probably within a large cloud of gas and dust around the sun, called an accretion disc from which Earth formed 4,500 million years ago.

Artist's conception of a protoplanetary disc

The Hadean Eon is not formally recognized, but it essentially marks the era before we have adequate record of significant solid rocks. The oldest dated zircons date from about 4,400 million years ago.

Earth was initially molten due to extreme volcanism and frequent collisions with other bodies. Eventually, the outer layer of the planet cooled to form a solid crust when water began accumulating in the atmosphere. The Moon formed soon afterwards, possibly as a result of the impact of a large planetoid with the Earth. Some of this object's mass merged with the Earth, significantly altering its internal composition, and a portion was ejected into space. Some of the material survived to form an orbiting moon. More recent potassium isotopic studies suggest that the Moon was formed by a smaller, high-energy, high-angular-momentum giant impact cleaving off a significant portion of the Earth. Outgassing and volcanic activity produced the primordial atmosphere.Condensing water vapor, augmented by ice delivered from comets, produced the oceans.

During the Hadean the Late Heavy Bombardment occurred (approximately 4,100 to 3,800 million years ago) during which a large number of impact craters are believed to have formed on the Moon, and by inference on Earth, Mercury, Venus and Mars as well.

Archean Eon

The Earth of the early Archean (4,000 to 2,500 million years ago) may have had a different tectonic style. During this time, the Earth's crust cooled enough that rocks and continental plates began to form. Some scientists think because the Earth was hotter, that plate tectonic activity was more vigorous than it is today, resulting in a much greater rate of recycling of crustal material. This may have prevented cratonisation and continent formation until the mantle cooled and convection slowed down. Others argue that the subcontinental lithospheric mantle is too buoyant to subduct and that the lack of Archean rocks is a function of erosion and subsequent tectonic events.

In contrast to the Proterozoic, Archean rocks are often heavily metamorphized deep-water sediments, such as graywackes, mudstones, volcanic sediments and banded iron formations. Greenstone belts are typical Archean formations, consisting of alternating high- and low-grade metamorphic rocks. The high-grade rocks were derived from volcanic island arcs, while the low-grade metamorphic rocks represent deep-sea sediments eroded from the neighboring island frogs and deposited in a forearc basin. In short, greenstone belts represent sutured protocontinents.

The Earth's magnetic field was established 3.5 billion years ago. The solar wind flux was about 100 times the value of the modern Sun, so the presence of the magnetic field helped prevent the planet's atmosphere from being stripped away, which is what likely happened to the atmosphere of Mars. However, the field strength was lower than at present and the magnetosphere was about half the modern radius.

Proterozoic Eon

The geologic record of the Proterozoic (2,500 to 541 million years ago) is more complete than that for the preceding Archean. In contrast to the deep-water deposits of the Archean, the Proterozoic features many strata that were laid down in extensive shallow epicontinental seas; furthermore, many of these rocks are less metamorphosed than Archean-age ones, and plenty are unaltered. Study of these rocks show that the eon featured massive, rapid continental accretion (unique to the Proterozoic), supercontinent cycles, and wholly modern orogenic activity. Roughly 750 million years ago, the earliest-known supercontinent Rodinia, began to break apart. The continents later recombined to form Pannotia, 600–540 Ma.

The first-known glaciations occurred during the Proterozoic, one began shortly after the beginning of the eon, while there were at least four during the Neoproterozoic, climaxing with the Snowball Earth of the Varangian glaciation.

Phanerozoic Eon

The Phanerozoic Eon is the current eon in the geologic timescale. It covers roughly 541 million years. During this period continents drifted about, eventually collected into a single landmass known as Pangea and then split up into the current continental land-masses.

The Phanerozoic is divided into three eras — the Paleozoic, the Mesozoic and the Ce-nozoic.

Paleozoic Era

The Paleozoic spanned from roughly 541 to 252 million years ago (Ma) and is subdi-vided into six geologic periods; from oldest to youngest they are the Cambrian, Ordovi-cian, Silurian, Devonian, Carboniferous and Permian. Geologically, the Paleozoic starts shortly after the breakup of a supercontinent called Pannotia and at the end of a global ice age. Throughout the early Paleozoic, the Earth's landmass was broken up into a substantial number of relatively small continents. Toward the end of the era the conti-nents gathered together into a supercontinent called Pangaea, which included most of the Earth's land area.

Cambrian Period

The Cambrian is a major division of the geologic timescale that begins about 541.0 ± 1.0 Ma. Cambrian continents are thought to have resulted from the breakup of a Neo-proterozoic supercontinent called Pannotia. The waters of the Cambrian period appear to have been widespread and shallow. Continental drift rates may have been anoma-lously high. Laurentia, Baltica and Siberia remained independent continents following the break-up of the supercontinent of Pannotia. Gondwana started to drift toward the South Pole. Panthalassa covered most of the southern hemisphere, and minor oceans included the Proto-Tethys Ocean, Iapetus Ocean and Khanty Ocean.

Ordovician Period

The Ordovician Period started at a major extinction event called the Cambrian-Ordovi-cian extinction events some time about 485.4 ± 1.9 Ma. During the Ordovician the south-ern continents were collected into a single continent called Gondwana. Gondwana start-ed the period in the equatorial latitudes and, as the period progressed, drifted toward the South Pole. Early in the Ordovician the continents Laurentia, Siberia and Baltica were still independent continents (since the break-up of the supercontinent Pannotia earli-er), but Baltica began to move toward Laurentia later in the period, causing the Iapetus Ocean to shrink between them. Also, Avalonia broke free from Gondwana and began to head north toward Laurentia. The Rheic Ocean was formed as a result of this. By the end of the period, Gondwana had neared or approached the pole and was largely glaciated.

The Ordovician came to a close in a series of extinction events that, taken together, comprise the second-largest of the five major extinction events in Earth's history in terms of percentage of genera that became extinct. The only larger one was the Permian-Triassic extinction event. The extinctions occurred approximately 447 to 444 million years ago and mark the boundary between the Ordovician and the following Silurian Period.

The most-commonly accepted theory is that these events were triggered by the onset of an ice age, in the Hirnantian faunal stage that ended the long, stable greenhouse conditions typical of the Ordovician. The ice age was probably not as long-lasting as once thought; study of oxygen isotopes in fossil brachiopods shows that it was probably no longer than 0.5 to 1.5 million years. The event was preceded by a fall in atmospheric carbon dioxide (from 7000ppm to 4400ppm) which selectively affected the shallow seas where most organisms lived. As the southern supercontinent Gondwana drifted over the South Pole, ice caps formed on it. Evidence of these ice caps have been detected in Upper Ordovician rock strata of North Africa and then-adjacent northeastern South America, which were south-polar locations at the time.

Silurian Period

The Silurian is a major division of the geologic timescale that started about 443.8 ± 1.5 Ma. During the Silurian, Gondwana continued a slow southward drift to high southern latitudes, but there is evidence that the Silurian ice caps were less extensive than those of the late Ordovician glaciation. The melting of ice caps and glaciers contributed to a rise in sea levels, recognizable from the fact that Silurian sediments overlie eroded Ordovician sediments, forming an unconformity. Other cratons and continent fragments drifted together near the equator, starting the formation of a second supercontinent known as Euramerica. The vast ocean of Panthalassa covered most of the northern hemisphere. Other minor oceans include Proto-Tethys, Paleo-Tethys, Rheic Ocean, a seaway of Iapetus Ocean (now in between Avalonia and Laurentia), and newly formed Ural Ocean.

Devonian Period

The Devonian spanned roughly from 419 to 359 Ma. The period was a time of great tectonic activity, as Laurasia and Gondwana drew closer together. The continent Euramerica (or Laurussia) was created in the early Devonian by the collision of Laurentia and Baltica, which rotated into the natural dry zone along the Tropic of Capricorn. In these near-deserts, the Old Red Sandstone sedimentary beds formed, made red by the oxidized iron (hematite) characteristic of drought conditions. Near the equator Pangaea began to consolidate from the plates containing North America and Europe, further raising the northern Appalachian Mountains and forming the Caledonian Mountains in Great Britain and Scandinavia. The southern continents remained tied together in the supercontinent of Gondwana. The remainder of modern Eurasia lay in the Northern Hemisphere. Sea levels were high worldwide, and much of the land lay submerged under shallow seas. The deep, enormous Panthalassa (the "universal ocean") covered the rest of the planet. Other minor oceans were Paleo-Tethys, Proto-Tethys, Rheic Ocean and Ural Ocean (which was closed during the collision with Siberia and Baltica).

Carboniferous Period

The Carboniferous extends from about 358.9 ± 0.4 to about 298.9 ± 0.15 Ma.

A global drop in sea level at the end of the Devonian reversed early in the Carboniferous; this created the widespread epicontinental seas and carbonate deposition of the Mississippian. There was also a drop in south polar temperatures; southern Gondwana was glaciated throughout the period, though it is uncertain if the ice sheets were a holdover from the Devonian or not. These conditions apparently had little effect in the deep tropics, where lush coal swamps flourished within 30 degrees of the northernmost glaciers. A mid-Carboniferous drop in sea-level precipitated a major marine extinction, one that hit crinoids and ammonites especially hard. This sea-level drop and the associated unconformity in North America separate the Mississippian Period from the Pennsylvanian period.

The Carboniferous was a time of active mountain building, as the supercontinent Pangea came together. The southern continents remained tied together in the supercontinent Gondwana, which collided with North America-Europe (Laurussia) along the present line of eastern North America. This continental collision resulted in the Hercynian orogeny in Europe, and the Alleghenian orogeny in North America; it also extended the newly uplifted Appalachians southwestward as the Ouachita Mountains. In the same time frame, much of present eastern Eurasian plate welded itself to Europe along the line of the Ural mountains. There were two major oceans in the Carboniferous the Panthalassa and Paleo-Tethys. Other minor oceans were shrinking and eventually closed the Rheic Ocean (closed by the assembly of South and North America), the small, shallow Ural Ocean (which was closed by the collision of Baltica, and Siberia continents, creating the Ural Mountains) and Proto-Tethys Ocean.

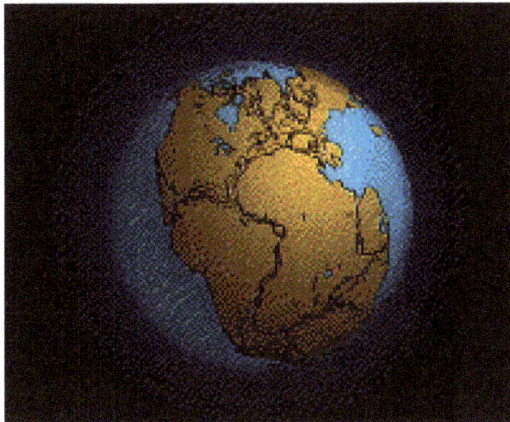

Pangaea separation animation

Permian Period

The Permian extends from about 298.9 ± 0.15 to 252.17 ± 0.06 Ma.

During the Permian all the Earth's major land masses, except portions of East Asia, were collected into a single supercontinent known as Pangaea. Pangaea straddled the equator and extended toward the poles, with a corresponding effect on ocean currents in the single great ocean (*Panthalassa*, the *universal sea*), and the Paleo-Tethys Ocean, a large ocean that was between Asia and Gondwana. The Cimmeria continent rifted away from Gondwana and drifted north to Laurasia, causing the Paleo-Tethys to shrink. A new ocean was growing on its southern end, the Tethys Ocean, an ocean that would dominate much of the Mesozoic Era. Large continental landmasses create climates with extreme variations of heat and cold ("continental climate") and monsoon conditions with highly seasonal rainfall patterns. Deserts seem to have been widespread on Pangaea.

Mesozoic Era

The Mesozoic extended roughly from 252 to 66 million years ago.

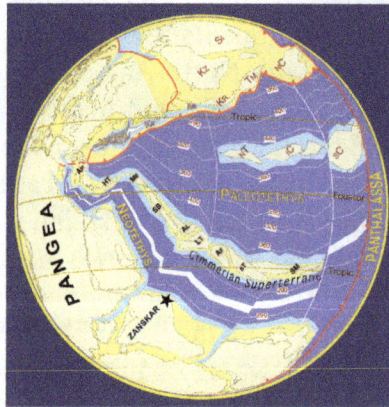

Plate tectonics- 249 million years ago

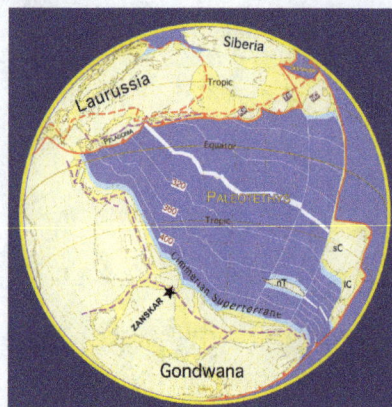

Plate tectonics- 290 million years ago

After the vigorous convergent plate mountain-building of the late Paleozoic, Mesozoic tectonic deformation was comparatively mild. Nevertheless, the era featured the dramatic rifting of the supercontinent Pangaea. Pangaea gradually split into a northern continent, Laurasia, and a southern continent, Gondwana. This created the passive continental margin that characterizes most of the Atlantic coastline (such as along the U.S. East Coast) today.

Triassic Period

The Triassic Period extends from about 252.17 ± 0.06 to 201.3 ± 0.2 Ma. During the Triassic, almost all the Earth's land mass was concentrated into a single supercontinent centered more or less on the equator, called Pangaea ("all the land"). This took the form of a giant "Pac-Man" with an east-facing "mouth" constituting the Tethys sea, a vast gulf that opened farther westward in the mid-Triassic, at the expense of the shrinking Paleo-Tethys Ocean, an ocean that existed during the Paleozoic.

The remainder was the world-ocean known as Panthalassa ("all the sea"). All the deep-ocean sediments laid down during the Triassic have disappeared through subduction of oceanic plates; thus, very little is known of the Triassic open ocean. The supercontinent Pangaea was rifting during the Triassic—especially late in the period—but had not yet separated. The first nonmarine sediments in the rift that marks the initial break-up of Pangea—which separated New Jersey from Morocco—are of Late Triassic age; in the U.S., these thick sediments comprise the Newark Supergroup. Because of the limited shoreline of one super-continental mass, Triassic marine deposits are globally relatively rare; despite their prominence in Western Europe, where the Triassic was first studied. In North America, for example, marine deposits are limited to a few exposures in the west. Thus Triassic stratigraphy is mostly based on organisms living in lagoons and hypersaline environments, such as *Estheria* crustaceans and terrestrial vertebrates.

Jurassic Period

The Jurassic Period extends from about 201.3 ± 0.2 to 145.0 Ma. During the early Jurassic, the supercontinent Pangaea broke up into the northern supercontinent Laurasia and the southern supercontinent Gondwana; the Gulf of Mexico opened in the new rift between North America and what is now Mexico's Yucatan Peninsula. The Jurassic North Atlantic Ocean was relatively narrow, while the South Atlantic did not open until the following Cretaceous Period, when Gondwana itself rifted apart. The Tethys Sea closed, and the Neotethys basin appeared. Climates were warm, with no evidence of glaciation. As in the Triassic, there was apparently no land near either pole, and no extensive ice caps existed. The Jurassic geological record is good in western Europe, where extensive marine sequences indicate a time when much of the continent was submerged under shallow tropical seas; famous locales include the Jurassic Coast World Heritage Site and the renowned late Jurassic *lagerstätten* of Holzmaden and Solnhofen. In contrast, the North American Jurassic record is the poorest of the

Mesozoic, with few outcrops at the surface. Though the epicontinental Sundance Sea left marine deposits in parts of the northern plains of the United States and Canada during the late Jurassic, most exposed sediments from this period are continental, such as the alluvial deposits of the Morrison Formation. The first of several massive batholiths were emplaced in the northern Cordillera beginning in the mid-Jurassic, marking the Nevadan orogeny. Important Jurassic exposures are also found in Russia, India, South America, Japan, Australasia and the United Kingdom.

Cretaceous Period

The Cretaceous Period extends from circa 145 million years ago to 66 million years ago.

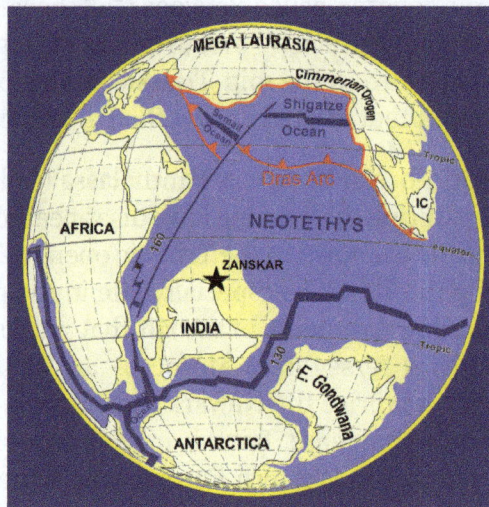

Plate tectonics- 100 Ma, Cretaceous period

During the Cretaceous, the late Paleozoic-early Mesozoic supercontinent of Pangaea completed its breakup into present day continents, although their positions were substantially different at the time. As the Atlantic Ocean widened, the convergent-margin orogenies that had begun during the Jurassic continued in the North American Cordillera, as the Nevadan orogeny was followed by the Sevier and Laramide orogenies. Though Gondwana was still intact in the beginning of the Cretaceous, Gondwana itself broke up as South America, Antarctica and Australia rifted away from Africa (though India and Madagascar remained attached to each other); thus, the South Atlantic and Indian Oceans were newly formed. Such active rifting lifted great undersea mountain chains along the welts, raising eustatic sea levels worldwide.

To the north of Africa the Tethys Sea continued to narrow. Broad shallow seas advanced across central North America (the Western Interior Seaway) and Europe, then receded late in the period, leaving thick marine deposits sandwiched between coal beds. At the peak of the Cretaceous transgression, one-third of Earth's present land area was submerged. The Cretaceous is justly famous for its chalk; indeed, more chalk formed in the

Cretaceous than in any other period in the Phanerozoic. Mid-ocean ridge activity—or rather, the circulation of seawater through the enlarged ridges—enriched the oceans in calcium; this made the oceans more saturated, as well as increased the bioavailability of the element for calcareous nanoplankton. These widespread carbonates and other sedimentary deposits make the Cretaceous rock record especially fine. Famous formations from North America include the rich marine fossils of Kansas's Smoky Hill Chalk Member and the terrestrial fauna of the late Cretaceous Hell Creek Formation. Other important Cretaceous exposures occur in Europe and China. In the area that is now India, massive lava beds called the Deccan Traps were laid down in the very late Cretaceous and early Paleocene.

Cenozoic Era

The Cenozoic Era covers the 66 million years since the Cretaceous–Paleogene extinction event up to and including the present day. By the end of the Mesozoic era, the continents had rifted into nearly their present form. Laurasia became North America and Eurasia, while Gondwana split into South America, Africa, Australia, Antarctica and the Indian subcontinent, which collided with the Asian plate. This impact gave rise to the Himalayas. The Tethys Sea, which had separated the northern continents from Africa and India, began to close up, forming the Mediterranean sea.

Paleogene Period

The Paleogene (alternatively Palaeogene) Period is a unit of geologic time that began 66 and ended 23.03 Ma and comprises the first part of the Cenozoic Era. This period consists of the Paleocene, Eocene and Oligocene Epochs.

Paleocene Epoch

The Paleocene, lasted from 66 million years ago to 56 million years ago.

In many ways, the Paleocene continued processes that had begun during the late Cretaceous Period. During the Paleocene, the continents continued to drift toward their present positions. Supercontinent Laurasia had not yet separated into three continents. Europe and Greenland were still connected. North America and Asia were still intermittently joined by a land bridge, while Greenland and North America were beginning to separate. The Laramide orogeny of the late Cretaceous continued to uplift the Rocky Mountains in the American west, which ended in the succeeding epoch. South and North America remained separated by equatorial seas (they joined during the Neogene); the components of the former southern supercontinent Gondwana continued to split apart, with Africa, South America, Antarctica and Australia pulling away from each other. Africa was heading north toward Europe, slowly closing the Tethys Ocean, and India began its migration to Asia that would lead to a tectonic collision and the formation of the Himalayas.

Eocene Epoch

During the Eocene (56 million years ago - 33.9 million years ago), the continents continued to drift toward their present positions. At the beginning of the period, Australia and Antarctica remained connected, and warm equatorial currents mixed with colder Antarctic waters, distributing the heat around the world and keeping global temperatures high. But when Australia split from the southern continent around 45 Ma, the warm equatorial currents were deflected away from Antarctica, and an isolated cold water channel developed between the two continents. The Antarctic region cooled down, and the ocean surrounding Antarctica began to freeze, sending cold water and ice floes north, reinforcing the cooling. The present pattern of ice ages began about 40 million years ago.

The northern supercontinent of Laurasia began to break up, as Europe, Greenland and North America drifted apart. In western North America, mountain building started in the Eocene, and huge lakes formed in the high flat basins among uplifts. In Europe, the Tethys Sea finally vanished, while the uplift of the Alps isolated its final remnant, the Mediterranean, and created another shallow sea with island archipelagos to the north. Though the North Atlantic was opening, a land connection appears to have remained between North America and Europe since the faunas of the two regions are very similar. India continued its journey away from Africa and began its collision with Asia, creating the Himalayan orogeny.

Oligocene Epoch

The Oligocene Epoch extends from about 34 million years ago to 23 million years ago. During the Oligocene the continents continued to drift toward their present positions.

Antarctica continued to become more isolated and finally developed a permanent ice cap. Mountain building in western North America continued, and the Alps started to rise in Europe as the African plate continued to push north into the Eurasian plate, isolating the remnants of Tethys Sea. A brief marine incursion marks the early Oligocene in Europe. There appears to have been a land bridge in the early Oligocene between North America and Europe since the faunas of the two regions are very similar. During the Oligocene, South America was finally detached from Antarctica and drifted north toward North America. It also allowed the Antarctic Circumpolar Current to flow, rapidly cooling the continent.

Neogene Period

The Neogene Period is a unit of geologic time starting 23.03 Ma. and ends at 2.588 Mya. The Neogene Period follows the Paleogene Period. The Neogene consists of the Miocene and Pliocene and is followed by the Quaternary Period.

Miocene Epoch

The Miocene extends from about 23.03 to 5.333 Ma.

During the Miocene continents continued to drift toward their present positions. Of the modern geologic features, only the land bridge between South America and North America was absent, the subduction zone along the Pacific Ocean margin of South America caused the rise of the Andes and the southward extension of the Meso-American peninsula. India continued to collide with Asia. The Tethys Seaway continued to shrink and then disappeared as Africa collided with Eurasia in the Turkish-Arabian region between 19 and 12 Ma (ICS 2004). Subsequent uplift of mountains in the western Mediterranean region and a global fall in sea levels combined to cause a temporary drying up of the Mediterranean Sea resulting in the Messinian salinity crisis near the end of the Miocene.

Pliocene Epoch

The Pliocene extends from 5.333 million years ago to 2.588 million years ago. During the Pliocene continents continued to drift toward their present positions, moving from positions possibly as far as 250 kilometres (155 mi) from their present locations to positions only 70 km from their current locations.

South America became linked to North America through the Isthmus of Panama during the Pliocene, bringing a nearly complete end to South America's distinctive marsupial faunas. The formation of the Isthmus had major consequences on global temperatures, since warm equatorial ocean currents were cut off and an Atlantic cooling cycle began, with cold Arctic and Antarctic waters dropping temperatures in the now-isolated Atlantic Ocean. Africa's collision with Europe formed the Mediterranean Sea, cutting off the remnants of the Tethys Ocean. Sea level changes exposed the land-bridge between Alaska and Asia. Near the end of the Pliocene, about 2.58 million years ago (the start of the Quaternary Period), the current ice age began. The polar regions have since undergone repeated cycles of glaciation and thaw, repeating every 40,000–100,000 years.

Quaternary Period

Pleistocene Epoch

The Pleistocene extends from 2.588 million years ago to 11,700 years before present. The modern continents were essentially at their present positions during the Pleistocene, the plates upon which they sit probably having moved no more than 100 kilometres (62 mi) relative to each other since the beginning of the period.

Holocene Epoch

The Holocene Epoch began approximately 11,700 calendar years before present and continues to the present. During the Holocene, continental motions have been less than a kilometer.

The last glacial period of the current ice age ended about 10,000 years ago. Ice melt caused world sea levels to rise about 35 metres (115 ft) in the early part of the Holocene. In addition, many areas above about 40 degrees north latitude had been depressed by the weight of the Pleistocene glaciers and rose as much as 180 metres (591 ft) over the late Pleistocene and Holocene, and are still rising today. The sea level rise and temporary land depression allowed temporary marine incursions into areas that are now far from the sea. Holocene marine fossils are known from Vermont, Quebec, Ontario and Michigan. Other than higher latitude temporary marine incursions associated with glacial depression, Holocene fossils are found primarily in lakebed, floodplain and cave deposits. Holocene marine deposits along low-latitude coastlines are rare because the rise in sea levels during the period exceeds any likely upthrusting of non-glacial origin. Post-glacial rebound in Scandinavia resulted in the emergence of coastal areas around the Baltic Sea, including much of Finland. The region continues to rise, still causing weak earthquakes across Northern Europe. The equivalent event in North America was the rebound of Hudson Bay, as it shrank from its larger, immediate post-glacial Tyrrell Sea phase, to near its present boundaries.

References

- Asimov, M. S.; Bosworth, Clifford Edmund (eds.). The Age of Achievement: A.D. 750 to the End of the Fifteenth Century : The Achievements. History of civilizations of Central Asia. pp. 211–214. ISBN 978-92-3-102719-2.

- Second J A (1986) Controversy in Victorian Geology: The Cambrian-Silurian Dispute Princeton University Press, 301pp, ISBN 0-691-02441-3

- Hooker, J.J., "Tertiary to Present: Paleocene", pp. 459-465, Vol. 5. of Selley, Richard C., L. Robin McCocks, and Ian R. Plimer, Encyclopedia of Geology, Oxford: Elsevier Limited, 2005. ISBN 0-12-636380-3

- Staff (March 4, 2010). "Oldest measurement of Earth's magnetic field reveals battle between Sun and Earth for our atmosphere". Physorg.news. Retrieved 2010-03-27.

Permissions

Index

www.ingramcontent.com/pod-product-compliance
Lightning Source LLC
Chambersburg PA
CBHW061930190326
41458CB00009B/2709

9 781635 491340